UNDER CONSTRUCTION

Building for Health in the EC Workplace

EF/92/16/EN

European Foundation
for the Improvement of
Living and Working Conditions

ZZ
EM860
92U51

UNDER CONSTRUCTION
Building for Health in the EC Workplace

by
Richard Wynne
Nadia Clarkin

Work Research Centre

Loughlinstown House, Shankill, Co. Dublin, Ireland
Tel: +353 1 282 68 88 Fax: +353 1 282 64 56
Telex: 30726 EURF EI

Cataloguing data can be found at the end of this publication

Luxembourg: Office for Official Publications of the European Communities, 1992

ISBN 92-826-4629-7

© European Foundation for the Improvement of Living and Working Conditions, 1992.

For rights of translation or reproduction, applications should be made to the Director, European Foundation for the Improvement of Living and Working Conditions, Loughlinstown House, Shankill, Co. Dublin, Ireland.

Printed in Ireland

TABLE OF CONTENTS

Preface

1. Why the Interest in Workplace Health ?   1

1.1 Introduction   1
1.2 Supra-national influences on health actions in the workplace   2
1.3 Problems in delivery of health actions in the workplace   3
1.4 The benefits of health actions   7
1.5 Opportunities for Developing Workplace Actions   9
1.6 Changes in the nature of work   13
1.7 So why the interest in Workplace Health ?   16

2. What Are We Talking About ?   17

2.1 What is Workplace Action for Health - What the objectives are   17
2.2 What is Workplace Action for Health - Examples of good practice   21

Germany - A Large Company and Integrated Health Policy   23
Italy: Health Promotion In The Ceramics industry   24
Portugal - Health, safety and quality of life in a local authority   27
Ireland - Social wellbeing in a rural area   30
The Netherlands - The "Healthier Work" project   32
The UK - The Welsh Heart Programme   35
Greece - the evolution of in-company health services   38
The Netherlands - Improving VDU Work   41
Spain - Applying health promotion principles   43
Sweden - A Practical Example of Improving Wellbeing at Work   46

2.3 So what are we talking about ?   48

3. The Playing Field - The Official Story.   51

3.1 Introduction   51
3.2 Legislative developments   52
3.3 The attitudes of key actors   69
3.4 So what is the Official Story ?   79

4. The Reality of Workplace Health   83

4.1 Questionnaire development   83
4.2 Sampling   84
4.3 Response rates   85
4.4 Demographics of the sample   86
4.5 Company Health Characteristics   88
4.6 Health actions in the workplace   93
4.7 So what is the reality of workplace health ?   99

5. What Actually Influences Workplace Health Actions ?   101

5.1 Introduction   101
5.2 Were there national differences in the samples   104
5.3 What was most important in explaining health activity   108
5.4 So what does influence workplace health actions ?   117

6. Establishing Health Actions   119

6.1 Why do organisations undertake health actions   119
6.2 Who participates in these health actions   125
6.3 Does involvement predict health activity   128
6.4 What are the barriers to health action   130
6.5 How are barriers to health action overcome   133
6.6 Overcoming barriers to action on workplace health   143
6.7 So how are workplace health actions established ?   145

7. What Does It All Mean ?   **147**

7.1 Introduction   147
7.2 Legislation   147
7.3 The views of the major actors   148
7.4 Lessons from the case studies   149
7.5 Lessons from the survey   151
7.6 The situation of SME's   153
7.7 National differences   154
7.8 Lessons on participation   155
7.9 Future prospects for workplace health promotion   156

8. What needs to happen ?   **161**

8.1 Introduction   161
Recommendation 1 - Information dissemination   164
Recommendation 2 - Research   166
Recommendation 3 - Training   168
Recommendation 4 - Legislation   170
Recommendation 5 - Enforcement of legislation   171
Recommendation 6 - Methodology for the establishment of   173
workplace health promotion
Recommendation 7 - Small to medium enterprises and   175
health promotion
Recommendation 8 - Health structures in the workplace   176
Recommendation 9 - Participation in workplace health actions   178

8.2 An action plan to encourage workplace health promotion   179

References   183

Appendices

Appendix 1   Reports published from the research programme

Appendix 2   Research instrument

List of Tables

Table 1.1	The eight national research agencies	v
Table 3.1	Summary of the interviews conducted	69
Table 4.1	Response rates in the different countries	85
Table 4.2	Health Problems of employees	89
Table 4.3	Health risks in the workplace - Findings from the Eurobarometer survey	91
Table 4.4	Proportion of activities which have health as a consideration	98
Table 5.1	Dependent and independent variables used in the analyses	102
Table 5.2	Economic sector by country	104
Table 5.3	Ownership by country	105
Table 5.4	Company size by country	106
Table 5.5	Levels of trade union membership by country	107
Table 5.6	Company health characteristics by country	107
Table 6.1	Prompting factors and the benefits of health actions	121
Table 6.2	Spontaneously reported benefits of establishing health actions	123
Table 6.3	Most important predictors of health activity	129
Table 6.4	Spontaneously reported problems in establishing health actions	131
Table 6.5	Job titles of respondents who completed the questionnaire	133
Table 6.6	Successful strategies used to overcome barriers to health actions	134
Table 8.1	Summary of the recommendations	161

List of Figures

Figure 4.1	Industrial sector of respondent organisations	86
Figure 4.2	Numbers employed in respondent organisations	87
Figure 4.3	Relative health priorities of the respondents	91
Figure 4.4	Prevalence of health screening activities	94
Figure 4.5	Prevalence of healthy behaviour activities	95
Figure 4.6	Prevalence of organisational interventions	95
Figure 4.7	Prevalence of safety/physical environment activities	96
Figure 4.8	Prevalence of social/welfare activities	96
Figure 6.1	Factors which prompted health actions	120
Figure 6.2	The benefits of health actions	120
Figure 6.3	Participation during the life cycle of health actions	126
Figure 6.4	Participation during the implementation stage	128
Figure 6.5	Problems in establishing health actions	132

Preface

A key part of the European Foundation for the Improvement of Living and Working Conditions (the Foundation) Four-year Rolling Programme for 1989-1992 concerns the promotion of health and safety. Three kinds of activity are undertaken by the Foundation in this regard - collecting EC-wide data on the state and evolution of living and working conditions, improving health and safety at the design stage of product development and promoting health and safety for all workers and citizens. With regard to the last initiative, the Foundation is interested in documenting and analysing actions for health in the workplace with:

> *".... particular emphasis on the processes which promote these initiatives, and the conditions for the transfer of innovation. Consideration will also be given to the economics, ethics and effectiveness of these actions."*

The Foundation's programme has a strong orientation to preventive strategies and worker involvement. It considers actions which have as their focus health and wellbeing, health behaviours and health conducive work environments rather than an exclusive focus on safety at work. It is also interested in developments in the workplace which reflect both worker and environmentally oriented approaches to promoting health at work. Finally, it investigates links which may exist between workplace health action and health promotion actions taking place in the surrounding community.

The aims of the research programme

This report synthesizes the results from the main phases of a three-year research programme. To date, three phases of the research programme have been completed:

- the development of a framework for investigating health issues in the workplace, partly through an international review of literature and policies for health at work;

- the writing of eight national reports which provide an overview of the workplace health situation in Germany, Greece, Ireland, Italy, the Netherlands, Portugal, Spain and the UK; and

- the conducting of detailed surveys on characteristics and mechanisms of development of workplace health initiatives in seven countries (all of the above except Portugal). In addition, in each of the seven countries up to four case studies of workplace health actions in multinational companies have been completed.

This report synthesises the material generated from all three phases of the research. It attempts to portray the ways in which workplace health initiatives become established, to give a flavour of good practice in the area, and to identify the factors which facilitate or constrain the development of health programmes and policies in the workplace.

The strengths and weaknesses of the research programme

Like any major research programme, the current research has its strengths and weaknesses. On the positive side, it provides the first major investigation into the nature of workplace health action in eight of the member states. In addition, the research programme provides much detail on how workplace health action is organised, with particular reference to initiating factors and levels of participation by different professional and representative groups in the workplace.

A particular strength of the research concerns its transnational character. It represents the first major effort to identify commonalities and differences between the practices prevalent in different parts of the European Community. It is possible to identify the different traditions of culture, geography and industrial history which operate in different regions in Europe, and which affect the ways in which workplace health is conducted.

The survey research carried out in each of the seven countries, though extensive, was not intended to be representative. Larger enterprises, public and private, were selected in two or three regions of each country, both to help with response rates and because in general company size is a factor in the implementation of health programmes.

Accordingly, this research serves the purpose of providing a systematic review of current practice; it aims to illustrate the context and mechanisms of workplace health action within each of the countries. It can be used as the basis for understanding process, and for deriving recommendations which are targeted towards generating increased workplace health activity. It is intended as a first step to opening the debate on workplace health promotion at a European level - hopefully in an informed way.

This report does not claim to be a *precis* or summary of the seventeen published reports upon which it is based. For the interested reader, detail on national findings and previous summaries of the research can be obtained from the Foundation or from the various national research agencies outlined in Table 1. (Appendix 1 contains a list of the main reports published to date from the research programme).

This report does, however, attempt to tell the story of workplace health action in Europe in an integrated manner. It draws on appropriate data and examples from each of the countries, and generates recommendations for the further development of workplace health action which can be effected by policy makers at EC and national level, and by the major actors within enterprises.

The results, conclusions and recommendations of the report were presented to an evaluation committee of the Foundation's Administrative Board on July 8th 1992. The group composed of representatives of employers organisations, trades unions and governments in the member states welcomed the new information and its presentation. Their debate emphasised the changing legislation and, in this, the contribution of initiatives at European level. The report opens a number of key issues that demand further discussion, notably on the balance of attention to work-related health conditions and promotion of general health in the workplace; on the ethics of workplace health initiatives, particularly concern about the misuse of health information on employees; and on the question of who should finance workplace action for health. In this European Year of Safety, Hygiene and Health Protection at Work, the Foundation is pleased to contribute a report which offers new information, documentation and analysis that will support future discussions on strategies to maintain and improve the health of the workforce.

Table 1. The eight research institutes

Country	Research Agency
Germany	Institute for Social and Health Research (IGES), Berlin
Greece	Ministry of Labour, Athens
Ireland	Work Research Centre, Dublin
Italy	CEDOC/University of Florence, Florence
Netherlands	TNO Institute of Preventive Health Care, Leiden
Portugal	National School of Public Health, Lisbon
Spain	Institut Municipal de la Salut, Barcelona
United Kingdom	Health Promotion Authority for Wales, Cardiff

Clive Purkiss

Director

1. Why the Interest in Workplace Health ?

This chapter deals with the background to the development of interest in workplace health actions in the past few years. The following areas are covered;

- *Supra-national influences on health promotion*
- *Problems of delivery of workplace health programmes*
- *Opportunities for action in workplace health*
- *The changing nature of the workplace*

1.1 Introduction

Workplace health action in Europe has a long history stretching back over a century. During that time the focus of workplace health actions has broadened and deepened from an initial concern with providing working conditions which had minimum safety requirements to a modern situation in which a wide range of potential threats to health and safety are systematically controlled within at least some workplaces. The role and extent of legislation and regulation has increased, and increasing awareness of the issues of workplace safety and health have led to greater amounts of activity in the area.

Despite these welcome long-term changes, there is a widespread perception that the extent of workplace health and safety coverage does not meet the demands of many working environments. In many countries the same kinds of problems are to be found: the delivery of services to the non-industrial sector, the lack of progress in instituting health and safety services in SMEs, the development of new threats to health and safety from emerging technologies and processes, the lack of training for workplace health and safety personnel, and the lack of effective intervention and monitoring tools are among the most commonly cited weaknesses of current occupational health and safety services.

This report synthesises the findings from a major 8 country, EC wide study of health actions in European workplaces. The research has much to say about how workplace health action is organised, what are the promoting and constraining factors on the development of workplace health action, and in particular it provides examples of new approaches to health in the workplace. The focus of the research is particularly on health, rather than safety activities, though the distinction between safety and health actions is often blurred in practice.

The report draws on research carried out in the Germany, Greece, Ireland, Italy, Netherlands, Portugal, Spain and the UK. However, in synthesising the research findings, the report does not attempt to provide a listing of national findings, rather it attempts to adopt a European perspective and to provide a framework for understanding the reasons for the current state of development of workplace health services in Europe.

Before describing the details of the research findings, it is necessary to chart in some detail the background to workplace health actions in terms of the major forces and issues which impinge on their development. Changes in legislation and regulation, problems in the delivery of workplace health services, new opportunities for promoting their development have all impacted on workplace health actions. The final part of this Chapter outlines the aims and structure of the research programme upon which this entire report is based.

1.2 Supra-national influences on health actions in the workplace

Many legislative and other factors have combined at an international level to influence the context in which health actions in the workplace take place either currently or in the near future. Among the most important of these are developments at the EC level, where the Social Charter and the Framework and related Directives are already setting a supportive background for the development of workplace health activities. Also of importance has been the activities of the ILO, particularly in relation to Convention 161 on occupational health services. These supra-national regulation instruments have helped set the agendas of national governments with regard to the focus and breadth of national legislation and regulations.

Also worthy of mention in this context are the activities of the WHO with regard to their Health For All by the year 2000 programme (HFA:2000), which has helped place health issues firmly on the workplace agenda.

More general factors which have emanated from larger scale developments in society include an increased awareness of health issues among the general population, the development of health promotion approaches to the presence of health problems and an increasing realisation that the workplace can be an appropriate location for the addressing general health problems, and not just the occupational and occupationally related diseases and illnesses. The broadening of approaches towards health and its definition have also contributed to producing a more favourable environment in which workplace health actions can take place. An increased concern about environmental problems has also placed attention on the possibilities of environmental interventions, in their broadest sense, being appropriate measures for the improvement of health in the population at large and in the workplace in particular.

1.3 Problems in delivery of health actions in the workplace

The potential for broadly based approaches to health in the workplace to develop faces a number of practical constraints. These constraints include problems of delivery, which already exist in relation to traditional occupational health services, but operate more powerfully in relation to broader based approaches. In addition, there are a number of other problems which relate particularly to the development of integrated approaches to health in the workplace.

1.3.1 Coverage of Occupational Health Services in Europe

Rantanen (1991), in his review of Occupational Health Services within the World Health Organisation European Region (covering thirty two countries) estimates that only 44% of workers within this region are covered by in plant or group occupational health services. A further 29% of workers are covered by occupational health services linked to primary health care and approximately 26% are not covered at all. Within these countries, coverage ranges from less than 10% of the working population to 100%.

He rightly points out that these figures are almost certainly over-estimates of the coverage of occupational health services since many services are less than complete in terms of the range of services that they offer, and also many would tend to be under staffed. Whereas there may be "legislative" coverage of the workforce, there may not be practical on the ground coverage in many instances. Allowing for these caveats, he estimates that no more than 30% of workers are covered by full occupational health services within the European region.

The coverage of these services should be seen against some statistics of work related health problems. It was estimated that there were 10.6 million accidents in 1982 within the WHO European region causing some twenty one thousand fatalities and approximately 650,000 cases of occupational disease per annum are reported.

Allied to these broad statistics, the knowledge that within certain countries at least, there are specific sectors of the work force which are not covered at all by occupational health services further points to the current limitations of occupational health services. In many cases these include the public sector and office based enterprises. Added to this failure to cover often quite large sectors of the working population are the well attested difficulties in delivering health care or occupational health care to small enterprises. This occurs not alone in the case of SMEs (enterprises with less than 500 workers) but particularly in enterprises where there are less than 100 workers, in which the majority of the working populations in many countries work. There are also well attested problems in delivering occupational health care to mobile work places (such as transport and distribution) and to the agriculture and maritime sectors.

The above account relates to provision of more or less traditional occupational health services to these sectors. The problems experienced in delivering occupational health services are multiplied when one takes a broader view of the delivery of more general health services to the working population. Whereas occupational health services have been under development for approximately a century, it is only in much more recent times that approaches involving health education and health promotion have begun to be seen in the workplace.

1.3.2 Lack of Training and Expertise

Another well attested problem in the delivery of health actions to work places concerns the lack of training and expertise of many individuals who are already involved and more particularly in the case of individuals who might wish to be involved. The research in this European Foundation research programme has indicated strongly that there are insufficient numbers of well trained people to deliver health activity within most workplaces in many countries. This problem is compounded by the lack of appropriate resources, infrastructure and personnel to meet these training needs in many countries.

1.3.3. Finance and Facilities

A lack of finance and suitable infrastructural facilities has also been cited as a barrier to the establishment of workplace health actions by many enterprises (especially SMEs). In Europe, the problems of financing health actions in the workplace appear to be considerable and would also appear to be more serious than those facing US enterprises. The difference between the US, where much of the cost of health care is carried by the employer, and Europe, where much of the cost of health care is borne by the State would appear to militate against the development of more comprehensive health programmes within European enterprises. Related to this problem is the issue of demonstrating the effectiveness of health actions in the workplace. If the value of initiating health actions in the work place can be demonstrated in financial terms many of the barriers or perceived barriers related to lack of finance may well be overcome.

Many enterprises also cite a lack of facilities as being a barrier to them implementing workplace health actions. However, it would appear that lack of facilities is rarely the problem that it might first appear (this theme will be returned to later on in the report). Companies who genuinely wish to implement health activities, can and do use facilities of other organisations be they state or the private sector to implement health actions.

1.3.4 The Prevailing Health Culture

The prevailing health culture within the society at large and also within workplaces would appear to act as a barrier against the development of new approaches to health in the workplace. The concept of health promotion in the workplace, for example, tends to be poorly understood. Much of the reason for this relates to the fact that health services in general are perceived to be treatment oriented in society at large, while in the workplace it appears to be particularly safety oriented. Both of these factors militate against development of genuine programmes of health promotion in the work place.

1.3.5 Tools for Implementation on Workplace Health Action

A cursory reading of the literature on workplace health action quickly reveals a major problem in implementing broader approaches to health in the workplace is a lack of suitable tools. The dearth of tools relates both to the implementation and evaluation of health programmes. In practice therefore, it seems many attempts at delivering broader health activity in the workplace fail because of the availability of only a limited range of tools which are seen to be inflexible and inappropriate in the context of specific workplaces, and crucially, at least in the eyes of management, the value of these programmes cannot be adequately demonstrated (this leaves aside the issue as to whether the broader approach to health in the workplace actually works).

1.3.6 The Role of Health Care Professionals in the Workplace.

Many of the existing health care professionals in workplaces have practiced within relatively established traditions, i.e. occupational medicine and safety in the workplace. There has been some criticisms of the rigidity or the narrowness of focus of the activities conducted within these areas. In addition, in some countries there appears to be considerable resistance from both of these groups of professionals to broadening their brief. The consequences of this resistance for the implementation of broader approaches to health in the workplace has been to retard its development. This issue is returned to in later chapters.

1.4 The benefits of health actions

There is also increasing evidence that the benefits of engaging in health actions are becoming much more widely perceived, despite some initial scepticism on behalf of organisations who felt that there were difficulties in demonstrating particularly the financial benefits of engaging in health actions. Much of this resistance would seem to have been overcome, even though this may not be general to all organisations - it is clear that many organisations who engage in health actions actively perceive considerable benefits from doing so.

The US experience (where admittedly a difference cost structure on health insurance operates in Europe) indicates that many major corporations have engaged in extensive health activities in the workplace, many of which can be regarded as residing within the field of health promotion. Examples of these corporations include Johnson and Johnson, IBM and others. Initial findings from these health programmes indicate that the benefits accrue in what might be termed soft returns, that is in health and wellbeing and morale of the workforce, has been widely reported. In addition many US corporations claim to be able to demonstrate significant financial benefits from their health programmes.

The findings from the current study also strongly support the view that benefits are there to be gained for enterprises who engage in health actions in the workplace. Large proportions of respondents to our survey indicated that they perceived both hard and soft benefits from the health actions that they engaged in. These included improvements in health, improvements in morale of the workforce, as well as the harder benefits of improvements in absenteeism, reduced accident rates and financial returns. It is noteworthy that these are claims that are coming directly from the organisations themselves. It is not a case of academic researchers demonstrating minute and measurable benefits.

Even though these companies demonstrated these benefits, it is not clear as to how they actually measure these. In some cases at least it is probable that they did not formally monitor and evaluate the benefits of these programmes. Nonetheless the impact of these programmes is sufficiently large for widespread perceptions of benefits to be perceived.

The findings of increased awareness of benefits would appear to be having a spill over effect into enterprises who have not engaged in health activities. Successful marketing of programmes such as Johnson and Johnson 'Live for Life' indicate that there is a readiness to take on board health actions in the workplace where benefits can be demonstrated. Similarly the experience of Health and Welfare Canada with their various health promotion prograr. mes in the workplace, have shown that even in times of recession, companies are still anxious to get involved in the field of health promotion.

The issue of assessing the benefits and costs of workplace health actions is a complicated but important aspect of debate within enterprises. This debate is of particular importance because of the emphasis which is often placed on the issue during the establishment (or not) of workplace health actions.

Any serious attempt to assess the costs and benefits of health programmes is hampered by a number of difficulties. In particular the limitations in cost-benefit assessment methods and the long term and intangible nature of many benefits conspire to make accurate assessment, at least in money terms, difficult if not impossible. Furthermore, the hidden costs of not engaging in health action (e.g. a workforce with poorer levels of health) are rarely taken into account by cost-benefit analyses.

The methods of cost-benefit accounting are sufficiently inaccurate as to cause problems in the assessment of benefits in many other areas of business life even where the benefits are more readily amenable to quantification and translation into money terms. For example, the financial benefits of investment in new technology, particularly in the office sector are difficult to demonstrate. Related to the limitations of methods is the issue of contamination of measurements by factors other than health programmes. For example, if a company implements a health programme and contemporaneously introduces technological change which improves working conditions, to what is improvement in health to be attributed ?

A further issue concerns the difficulties of translating intangible benefits such as health improvements into measures of direct concern to enterprises. Though such indicators as absenteeism, accident rates and productivity can undoubtedly be affected by health programmes, the link between health improvement and these indicators is difficult to quantify. Finally, the long term nature of the payoffs of many health programmes (e.g. fitness programmes) makes the issue of benefit assessment even more difficult. (For a fuller discussion of methods for the assessment of costs and benefits, see Sloan et al, 1987).

These limitations of cost benefit methods have lead many to the conclusion that it is impossible to adequately demonstrate benefits on financial terms. An alternative approach is to emphasise the health programmes are intrinsically worth doing, regardless of the above mentioned difficulties. In essence, this view holds that the intangible benefits which accrue, for example improvements in morale, subjective wellbeing and organisational climate are sufficient to justify engaging in health programmes. In addition, many company cultures provide for health and welfare programmes as part of their mission. Some of the US multinationals and some of the southern European countries are good examples of this kind of justification.

1.5 Opportunities for Developing Workplace Actions

The workplace is increasingly been seen as an important and legitimate arena for action on health. This perception has grown for a number of reasons. In health circles it has been recognised that there are limits to the possibilities for preventive approaches through the medium of existing health care facilities, and that a move towards community- based initiatives is likely to be more successful. In particular the workplace has been seen as an arena where groups of people spend large amounts of time and are therefore an ideal target group.

Allied to this reasoning has been the realisation that the workplace can have a major role to play in influencing health and wellbeing of the worker. Leaving aside the obvious potential of workplaces to negatively influence health (this is the traditional domain of health and safety practice), the effects of the workplace on wellbeing have become clearer in the past twenty years or so. At it's crudest level it is evident that the workplace can bestow significant benefits in terms of wellbeing - the contrasts in wellbeing between the unemployed and the employed attest to this. But it has also become evident that the workplace can have significant negative effects of physical and mental wellbeing. Psychosocial factors such as shiftwork, work overload, machine pacing of work speed and poor relationships at work have all been shown to lower physical and psychological wellbeing. Perhaps a more fundamental point about work concerns the socialisation process which takes place in all workplaces. It has been shown that younger workers take many of their values and beliefs from the work setting.

For these reasons - the presence of large numbers of people for long periods of time, the positive and negative effects of the workplace on health and wellbeing - the workplace has been seen as a potentially promising arena for the promotion of health. Of course there are also other reasons related more to the context in which enterprises operate and the internal process of organisations which combine to increase the chances of health promotion taking place in the workplace. Some of these are outlined below.

1.5.1 Developments in legislation

Primary among these is the fact that legislation within the EC member states has become and is in the process of becoming much more supportive to broad interpretations of what constitutes workplace health activity. As outlined in Chapter 3, the implementation of the Framework Directive and other related Directives which will ultimately be enacted in national legislation will effectively create a higher floor under the level of health activities which take place in the workplace. The Framework Directive thus provides an example of the power of good legislation to effect change, at least at the legislative level.

In addition, these Directives introduce concepts which are new to many countries legislation which could be seen to be beyond the brief of more traditional health services in the workplace. In particular, aspects of stress in the workplace, the rights to participation in workplace health activity given to workers, and the provisions for training in workplace health and safety lay down conditions which when fully implemented are likely to lead to broader based approaches to health in the workplace.

1.5.2 Organisational change processes

Another factor which may be seen as an opportunity to promote health in the workplace concerns the rate and nature of organisational change which is affecting all sectors of European economy. It is well recognised within economic circles that companies survival depends on their becoming more flexible and dynamic. In effect, they must be prepared to change their structures and processes at an apparently ever increasing rate. If only from general organisational theory, these organisational changes offer the opportunity to introduce health programmes or health considerations into the process of organisational change. In practice, there are many examples of using organisational change to promote health and wellbeing among the workforce, and some good examples of these are outlined in the next chapter.

1.5.3 Changes in health culture

Notwithstanding the comments made about the prevailing health culture in society and in workplaces, there is evidence that the concepts of health and health care delivery are changing. In particular, there is a general move away from treatment oriented culture towards a more prevention based approach to health (at least in policy circles). As this new idea diffuses throughout society and workplaces, it provides increased opportunity for more broadly based approaches to health in the workplace to take place.

1.5.4 The attitudes of the social partners

The social partners (i.e. government, employers and trade unions) would appear to be showing a greater interest in health in the workplace. From a governmental point of view, whether under the influence of EC Directives or not, there is evidence that there are changes in legislation which will make the opportunities for health promotion in the workplace more positive. From the employers point of view, there would also appear to be increasing concern about health and safety in the workplaces, particularly as more and more of the companies realise the extent of problems like absenteeism and the inherent and potential dangers in many new workplace processes. From the trade unions point of view, there is some refreshing evidence that health in the workplace is no longer seen as matter for monetary compensation, rather it is seen as an issue which should be dealt with in a preventive manner.

1.5.5 Recognition of benefits

There is also a body of evidence which will be presented in this report which indicates that within companies where broader based approaches to health have been implemented that there is increasing recognition of benefits to be gained from engaging in such activities (particularly this would appear to be the case in US multinationals). At the most basic level, many companies having once embarked upon a health programme, do not discontinue their activities. Furthermore, there is evidence which indicates that management recognise that certain benefits both in health and non-health terms can be gained from engaging in such activities. (This appears to be the case despite a lack of evaluation instruments). As recognition and awareness of the benefits of health activities grows, it is likely that an increasing constituency of support for the implementation of such activities will be created.

1.5.6 Development of tools

Despite the perceived lack of suitable tools for the development of broadly based health actions, there are encouraging developments in this area, both within the academic sphere and more importantly the sphere of practice. In particular, the efforts to integrate health issues into work design and organisation by researchers such as Karasek and Theorell (1990) offer particular hope. Similarly, some of the developments in Holland and Germany (e.g. Grundemann, 1990, Grundemann and Ellis, 1991; Hauss, 1990, 1991) provide practical models of broadly based approaches to workplace health action.

1.6 Changes in the nature of work

Over the past decade there has been an increasing rate of change in the nature of work undertaken by the working population in Europe. There has been an increasing flight from the land (a reduction in the numbers working directly in the Agriculture Sector), a reduction in the numbers working in the Manufacturing Sector, a corresponding increase in the numbers working in the Services Sector and a significant ageing of the working population across much of Western Europe.

A further demographic change concerns what might be termed the 'feminisation' of the workplace. Across Europe, there has been a significant increase in the numbers of women working in the workplace in the past twenty years. Currently 39% of the EC workforce are female, with rate varying between 31% (Spain) and 45% (Denmark).

The increasing labour force participation by women has led to some changes in the ways in which work is organised, for example the introduction of flexitime has been associated with increasing participation of women in the labour force as has the increase in numbers engaged in part-time work. A recent survey carried out by the European Foundation on working conditions in the EC (Paoli, 1991) indicates some of the potential health risks which are associated with women. These include having less autonomy at work and having less input into how they organise their work. (Both of these factors are associated with stress in the workplace). Other findings suggest that women generally are over-represented in a high risk group characterised by high levels of organisational constraints on their working conditions.

Allied to these major demographic changes has been a change in nature of work within these Sectors. For example within the Manufacturing Sector, there has been a change from labour intensive working towards increasingly automated forms of manufacturing. Within the Services Sector there has been a move away from lower grade clerical work towards information technology based work.

The ageing of the European workforce in the more developed economies has begun to give rise to different health concerns in enterprises. The health of the workforce will become more of an issue for employers, as they seek to maintain workers in work and to reduce the incidence of chronic illness. Traditional approaches to health and safety, with their emphasis on workplace based hazards, will need to expand to take account of the problems of an ageing workforce. The strengths of health promotion, with its emphasis on health maintenance and improvement offer promise to employers wishing to combat the health problems of an ageing workforce.

Much of the legislation currently on the statute books on Europe and many of the occupational health services currently in existence are framed and set up in relation to workplaces which have substantially changed in the last decades. For example, within the Manufacturing Sector the concentration of occupational medical services on various toxic hazards in the workplace would appear to be increasingly inappropriate given the changes in the nature of manufacturing which have generated new hazards and imported many of the psychosocial hazards more usually associated with office work. A related point concerns the low levels of coverage of occupational medical services within the service sector, where often for legislative reasons occupational medical services in this sector are not widespread. It is important that there is not direct importation of traditional models of occupational health and a focus on traditional issues of occupational health, because by and large the threats to the safety and health of a very different nature and those experiences in the manufacturing sector.

Against this background it is important that as work becomes increasing less manual and increasingly more sedentary and information based that new models of health delivery in the workplace are generated. In this context health promotion in the workplace provides particularly promising methods of approach. The concentration on prevention and on dealing with the entire individual and their health rather than focusing solely on workplace based hazards is particularly appropriate in the context of service industries and increasingly so for the new manufacturing industries. Though this realisation may be taking root slowly it is a factor which will increasingly come to influence workplace health practice in the future.

1.7 So why the interest in Workplace Health ?

Broadly speaking there are two kinds of factors which have increased interest in workplace health promotion. These include the push factors (factors which operate external to the enterprise) of legislative changes both at a supranational and national level. The pull factors (intra-enterprise factors) on the other hand include the benefits to be gained from engaging in health actions and the increased opportunities for actually engaging in workplace health actions. Perhaps the principal reason is that workplace health works - there is an increasing number of companies who see the benefits of engaging on workplace health activity.

2. What Are We Talking About?

What is workplace health action ? In this chapter the approach to workplace health actions adopted in this research is described. This approach takes account of more 'traditional' approaches to workplace health which stem from mainstream health and safety services in the workplace, but is essentially concerned with newer and more innovative approaches to workplace health. Ten examples of good practice drawn from the many case studies conducted throughout Europe as part of the research are used to illustrate these new approaches.

2.1 What is Workplace Action for Health ? - What the objectives are

Current thinking recognises that health cannot be defined solely in terms of absence of disease, but should also incorporate wellbeing in its physical, psychological and social manifestation. These elements of wellbeing should have a positive focus. In addition, health is currently seen as a dynamic condition, in which the individual continuously adapts to the challenges (physical, biological, psychological and social) of a constantly changing environment.

However, many of the health actions which take place in workplaces are largely informed by definitions of health which focus on disease and disease prevention. Typically they are concerned with identifying the presence of agents in the workplace environment which may cause disease, and with the control of these hazardous agents. They focus on prevention of occupational disease and health protection through safety actions rather than on actively promoting good health. A feature of these approaches is that they often deal with potential health problems in a *post hoc* manner, in the sense that they only react to health threats which are perceived, rather than identifying improved health as a goal to be pursued independently of the presence of health threatening agents in the workplace.

The types of workplace health activity with which this research is largely concerned move beyond this preventive and protective approach. They incorporate elements of these approaches but they also adopt many of the strategies of health promotion.

What do we mean by health promotion in this context ? Defining health promotion in an exact and useful way is not an easy task. Indeed, many people working in the field have difficulty in distinguishing health promotion from health education and disease prevention. Part of the problem of defining health promotion relates to a tension between some of the formal definitions of health promotion, which put forward an idealistic and comprehensive meaning, and the common practice of what is described as health promotion by practitioners in the area. This common practice almost invariably involves workplace health actions which are limited in scope, and which tend to be directed exclusively at the individual, and which perhaps more properly fall under the banner of disease prevention strategies.

Anderson (1987) suggests that it may be more productive to view health promotion in terms of its principles of approach rather than in terms of specific activities. He outlines five such principles based on the ecological view of health proposed by the World Health Organisation (WHO, 1984). These are:

- *"Health Promotion involves the population as a whole in the context of their everyday life, rather than focusing on people at risk for specific diseases.....*

- *Health Promotion is directed towards action on the determinants or causes of health.....*

- *Health Promotion combines diverse, but complementary, methods or approaches, including communication, education, legislation, organisational change and community development.....*

- *Health Promotion aims particularly at effective and concrete public participation.....*

- *Health Promotion is basically an activity in the social, political and welfare fields, and not primarily a medical activity.....*"

Implicit in these principles is a continuum of activities ranging from prevention activities and education activities which lead to an emphasis on positive health to the conduct of activities which actively seek to promote health.

There is a good deal of terminological confusion in this area. There is, for example, a large body of work in the US and increasingly in the UK which describes so-called workplace health promotion programmes. These health promotion initiatives however, are often solely or principally concerned with modifying the behaviour of the individual, and pay scant regard to the environmental determinants of health. Though they may seek to promote the positive health of the individual, the fact that they do not meet all of the principles of approach outlined above could be interpreted as a failure to qualify as 'true' examples of workplace health promotion. For present purposes, however, such initiatives are included within the meaning of the term 'health promotion', as the practice of health promotion in the workplace rarely meets all of five principles of approach.

The WHO's five principles of approach to health promotion have been altered to tailor them for use in the workplace by Wynne (1989). In doing so, the first of the WHO principles was modified most, to ensure that the accurate targeting of workplace health actions was possible. The modified principles state that workplace health action:

- Can be applied across all groups in the workforce;

- Is directed at the underlying causes of ill health;

- Combines diverse methods of approach;

- Aims at effective worker participation; and

- Is not primarily a medical activity, but should be part of work organisation and working conditions.

These principles were used by the researchers in the eight countries in their descriptions of the case studies which were undertaken. Their intention is to enable the characterisation of new approaches to the issue of workplace health. It should be noted that these criteria of innovation are not absolute and serve only as guidelines, as it is not possible to accurately quantify many of the concepts outlined in them. Furthermore, it is not always possible to define whether a health action is primarily a workplace health action. Many valuable health actions take place primarily in arenas other than the workplace, though they may contain a workplace component.

It should be made clear that the above working definition includes what might be viewed as public health measures as well as measures derived from occupational health. Such actions as health education, providing healthy eating choices in canteens and providing exercise facilities stem directly from the public health tradition.

In the current research programme workplace actions for health have been operationally defined in the questionnaire used in the survey. In all, thirty different potential health actions (see Chapter 4) were defined under the headings of:

- Health screening;

- Actions which promote healthy behaviour;

- Social and welfare services;

- Organisational interventions; and

- Safety and physical work environment actions.

While the survey provides useful data about the prevalence of these activities, the case studies (outlined below) provide insight into how actions such as these can be combined to address real problems deriving measures from both the public and occupational health traditions.

A note on 'Innovation'

In the previous phases of the research, a certain emphasis was placed on the term 'innovative workplace health actions'. At that time 'innovative' was used to refer to workplace health initiatives which met all or nearly all of the criteria of assessment derived from the WHO principles of approach. However, the practice of health actions in European workplaces as revealed through the research demonstrated that truly 'innovative' health actions essentially do not exist. Many health actions are both necessary and useful without meeting the criteria of innovation. In some settings, the existence of a health action at all is highly innovative.

Within the present report the use of the term 'innovative' has been dropped, as it served only to place a problematic dimension of evaluation above the spirit of many health actions. In the present report, the issue of whether a health action meets the principles of approach to health promotion is dealt with only in terms of description. The national and organisational contexts within which health actions are taking place provide far superior metrics for the evaluation of the extent to which an action is innovative or not.

2.2 What is Workplace Action for Health ? - Examples of good practice

As can be deduced from the above account there are difficulties in theoretically describing workplace health actions in a satisfactory manner. However, one of the principal strengths of this research programme is that it has provided a wealth of information and data from each of the eight countries about the practice of workplace health. In this section some of the material from the many case studies conducted by the national researchers is presented for purposes of illustrating good practice, and in particular for purposes of illuminating the comprehensive and new approaches to health in the workplace which currently take place in Europe.

The case studies were selected on the basis that they provide a wide range of activities which take place against very different backgrounds. Each of the case studies outlined below is presented in a stand alone format which describes the health initiative in question. These are followed by commentaries which outline the distinctive features of the case study, and which point to differences between the case and some of the more traditional approaches to workplace health.

The first of these case studies comes from Germany and is concerned with a new approach to organising health activities in a large company. The second case study comes from Italy, and provides a fascinating insight into how a traditional occupational health problem can be dealt with in a small and distributed industry. The third case study comes from Portugal, and describes a programme for the improvement of health safety and quality of life in a local authority. It gives an account of how health promotion can be set up even in the most unpromising of circumstances. The fourth case study comes from a small town in Ireland, and illustrates how a flexible approach to health in the workplace can have major benefits for both workforce and the community in addition to addressing some of the problems which emerged when a large industry sets up in a small town. The fifth case study comes from the Netherlands and is concerned with a systematic approach to work stress, an emerging problem in many workplaces (at least in Northern Europe). The sixth case study comes from the UK and illustrates how broader health promotion programmes can be integrated into the workplace. The seventh case study comes from Greece where a long standing interest in worker welfare has manifested itself in an almost unique mix of programmes in a cement company. The next case study looks at a Dutch method for dealing with another emerging workplace health hazard - VDUs. The ninth case study addresses the important issue of the integration of health policy into wider company operations in the Barcelona municipal transport company.

The remaining case studies are taken from the general literature and provide an illustration of some more limited interpretations of workplace health promotion. While the case study taken from Sweden took place in a small company, and demonstrates the possibilities of health promotion for small companies, the remaining examples taken from the literature concern actions which are much more limited in scope, and are perhaps more typical of what is called workplace health promotion than the cases described as part of this research programme.

Germany - A Large Company and Integrated Health Policy

Health promotion in this company, goes beyond individual behaviour modification, and is seen as a comprehensive task which has little to do with strictly health related measures but which operates using the same kinds of industrial relations procedures used for recruiting, wages, designing of the firm's social relations etc. Health promotion is viewed in the context of other industrial relations issues among which health policy may not be of the highest importance. On the other hand, strong emphasis is placed on improving the efficiency of the OHS system by innovative measures. This is achieved by combining traditional OHS measures with those that are really innovative.

The firm operates a well designed scheme for improved coverage of OHS activities in addition to monitoring the health and risk-status of the employees. This procedure is based on sickness fund data and data originating from the firm's own reporting system. For example, long-term analyses of these data enables the firm, for instance, to generate risk profiles for specific chemicals and, if necessary to direct special measures to such a risk situation. Another aspect of the traditional OHS activities involves the participation of occupational physicians in designing new working conditions whenever new products are being developed. OHS in this enterprise must assess the potential health impact of new products before beginning production. This kind of decisive participation in health related aspects of production is often called for by West German OHS activists but has only rarely materialized.

Innovative activities include:

- Providing an elaborate stress management course for foremen. This course aims to reduce heart disease incidence by providing social support, peer validation of feelings and producing solutions to stress problems within the company.

- Another target group (older employees, mostly women, upholsterers) partake in twice daily gymnastics sessions. This is based on biofeedback procedures which adapt the exercise to each persons fitness level.

Commentary

This case study from a large German company, in some ways, presents an ideal example of the organisation of health activities in the workplace. In particular, its integration not alone of traditional occupational health activities but also of health promotion activities into the ongoing internal negotiations within the company are striking as is pointed out by the author (Hauss, 1990). The situation of effective integration of health issues into company planning regarding new products and processes is also remarkable. Another notable feature of this case is the extent to which data are used for the identification of potential workplace hazards and also for the accurate targeting of health programmes in the workplace. Again this situation has often been cited as being the ideal, but is rarely seen in practice.

Another feature of this case study is that the impetus for developing such an integrated programme came largely from within the company, thereby avoiding many of the problems associated with external consultants providing programmes which may not be targeted to needs of the company. As a result of this internal development, there are relatively few programmes provided by the company which focus on the individual behaviour modification. These kinds of programmes are typically seen as the responsibility of the individual.

Italy: Health Promotion In The Ceramics industry

This case study describes the intervention of the public occupational medical service in the ceramics industry of Emilia-Romagna, which consists of 270 companies with 28,000 employees.

Health protection problems in this industry were identified using a territorial map of hazards which provided a starting-point for the intervention, with a prevention plan targeted at one or more priority objectives present in different parts of the area. The hazard mapping was carried out by means of site investigations and environmental and health surveys. These investigations were not confined to workplaces, but were extended to community living spaces such as schools and houses where families had installed ovens for drying tiles which they had hand decorated. Investigations showed that lead pollution was widespread in both the workplace and the home, e.g. meals were eaten with hands still stained by paint containing lead.

The "targeted prevention plan" was linked to the gravity of the problem: 21% of the 6,000 workers exposed to lead showed signs of metabolic alterations. The extent of the problem had also been underestimated. The number of reports of lead poisoning changed from a few units per year to 254 in the first year of activity. Furthermore, in many factories workers took chelating agents or drank milk supplied by the employers to guard against lead poisoning. These were not only useless, but damaging both organically and because they gave workers a false sense of protection. Finally, the usual workplace health checks were pointless because they did not test for lead poisoning.

The initiative was designed to raise awareness of risks among workers, trade unions and employers, health authorities, doctors and the public at large. Pathologies were not treated as a series of isolated "cases" which the doctor was obliged to report, leaving it to the insurance sector to pick up the tab. Instead, they were analysed in collaboration with the workers themselves.

Updating the scientific literature

The hazards mapping contributed new material to the scientific literature, which had held that the only ceramics related problems were silicosis and lead poisoning. Other associated risks such as ionising radiation, allergic pathologies, diseases associated with pregnancy, chromium, postures, workloads, the micro-climate and noise were added to the literature.

Information as the basis of health promotion initiatives.

Information was a central feature in this programme. Hazard mapping provided the basis of a report to employers and also became a topic for meetings with workers. At these meetings findings were explained and solutions were sought. The multidisciplinary team from the OHS provided inputs, the effectiveness of which was checked by workers in relation to their own experience. Workers meetings were followed by meetings with management, trade unions and company doctors.

From the workplace to the community

In the first year of activity meetings were organised in factories and many community settings to disseminate information particularly to those who had suffered from hazards produced in the factories. However, the overall aim was to involve the entire population. This was achieved through workers meetings and also through meetings held with teachers and students, who were seen as people who could promote the development of a new health culture. Teachers used the practical approach of including information on lead pollution into the local schools curriculum. This created a lot of interest among teachers and students, who generated educational materials concerning environmental hazards and behaviour which needed to be modified.

Other results of the initiative

Another aim of the initiative was to change competence as well as knowledge and culture. Training courses were organised for workers, trade unions, employers, company doctors, teachers and the general public giving information on risks and hazards caused by the ceramics industry.

Over ten years, this combined approach led to radical improvements. Data from 1973 and 1983 show that the percentage of workers with alterations in their ALA-U levels fell from 29.9% to 0.8%.

The programme laid the groundwork for legislative initiatives in support of companies, including financial plans for the installation of purification equipment on a shared basis, and other initiatives relating to reducing environmental impacts. This led to the industry being prepared to use health and environmental protection techniques, while also encouraging local authorities to identify a planning and monitoring procedure.

Commentary

This case study provides a good example of a health programme which covers the entire sector of an industry and not just individual companies. In particular, it stresses some of the major problems associated with an industry which operates to some degree in the manner of a cottage industry. As with the first case study, an important feature of this case study is the use of data collection to enable a targeted programme to be developed. In addition, it demonstrates the effectiveness of having both a workplace and community based intervention approach. Even though some of the main activities of the programme are concerned with prevention and treatment, the combination of community and worksite based initiatives designed on the basis of accurate data are important channels of approach to problems which have both a workplace and environmental component.

Because of the high levels of ignorance relating to the effects of hazards among the companies, the workforce and indeed the occupational health service within the region, a major strategy of this initiative was the provision of information to all concerned. The level of cooperation between the institutions and services, companies, trade unions and the communities enabled this initiative to effectively address this knowledge gap. The success of this major programme is illustrated by the fact that the number of workers reporting lead poisoning dropped from 29.9% to 0.8% over the course of ten years.

Portugal - Health, safety and quality of life in a local authority

The number of workers in the local authority is about 1700, of whom about half are unskilled labour. The absenteeism rate is of the order of 11%. Accident levels are high, accidents at work being both frequent and serious. In addition, workers are poorly paid and have low self-esteem. Their work is not valued by society. Some of the workers are of African origin from Cape Verde and live in temporary accommodation.

Anticipating future legislative changes, the local authority set up an occupational health service, drawing on the experience of reorganizing the health services of TAP - Air Portugal.

With the reorganization of the local authority in 1987, the activities of and responsibility for the local authority's personnel cultural and sports centre passed to the personnel department. It is in this context that the occupational health project was born in 1989, and this involved the integration of health care and the negotiation of an insurance package (including sickness, life and accidents at work). The process of negotiation, both with the council executive and the insurers, was not easy. An attempt was made from the outset to involve structures representing the workforce.

The explicit objectives of this project are, among others, the promotion of workers' health and well-being, improvement of their living and working conditions and equitable treatment from the point of view of health and illness.

Action for safety and health at work

One of the most notable features of this project is the existence of a real occupational health team, consisting of more than 20 persons for some 9000 users (workers and their immediate families). One of the main areas of the OHS is primary health care. Other areas are: occupational medicine, industrial psychology, physiology and ergonomics, industrial hygiene, safety, epidemiology and biostatistics.

Joint programmes in various areas (some still at the planning stage) include: the prevention and early diagnosis of certain cancers, action on problems related to alcohol consumption, the setting up of a committee on working conditions, safety and health, education for health and for life, the management of occupational stress, ergonomic activities (eg working clothing), the prevention of back pain, break time and ongoing keep-fit sessions, the prevention of work-related skin diseases, scanning for cardiovascular diseases, etc. Priority has also been given to a programme for the prevention and early diagnosis of cancer of the neck of the uterus and breast, given that one out of every four of the authority's workers is female.

An important feature of this programme is that examinations are carried out on a day on which the women do not work, and which has been termed a "health day". The object of this is to de-dramatize screening for cancer, emphasizing health and not the disease.

Action on alcohol problems is felt to be technically the most complex programme, involving a large proportion of the human resources of the OHS. The programme began with a wide-ranging campaign of awareness and explanation on alcohol, work and health. As a teaching exercise, equipment to measure blood-alcohol levels after meals was set up in the Primary Health Care Office next to the main canteen, staffed and directed by nursing personnel.

An accident prevention programme has also been instituted, the main objective of which is to control risk factors and, specifically, to change attitudes and behaviour. In addition, a new participatory research methodology on the prevention of accidents at work as well as rehabilitation on the job have also been introduced. Until recently, accident victims were directed to the personnel department and from there to the insurers. The system took responsibility away from the management, which was almost only concerned with the consequences of accidents in terms of incapacity and absenteeism. New systems of investigation and rehabilitation on the job have also been introduced as part of the programme.

Another important feature of the programme has seen the setting up of a network of safety workers with one worker at each place of work who will also act as an agent for the promotion of workers' health and as an agent for the entire project.

Results to date, indicate that there are high levels of commitment to the project on the part of both management and workers' committees. As far as the employees in general are concerned involvement varies depending on the type of programme. For example, the involvement of working women in the cancer programme has been of the order of 60% to 75%. Involvement in the alcohol programme is less, at around 50%. Involvement in the workers' safety network is high, and there is also active participation by management in the stage of investigating the causes of accidents and incidents.

The costs and benefits of the programme are also beginning to emerge in terms of reduced seriousness of accidents (though not numbers). In addition, there has been a large decrease in working days lost from 1278 to 296, i.e. the number of days lost has fallen in the proportion 4.3 to 1.

There are also innovative aspects to the programme in the Portuguese context. Programmes such as the cancer prevention programme are fairly infrequent in Portuguese workplaces. The alcohol programme is likewise innovative in the present context.

Commentary

Perhaps the most interesting aspect of this case study concerns the unpromising background of a local authority operating in a country with no developed tradition of occupational health services, and a medical culture which was predominantly curative in nature. While the health actions themselves are not especially innovative, the fact that they have been combined on the basis of a preventive philosophy is of importance. The case study is also of interest in the way that insurers and the OHS services have combined to develop the programme. Though this course of action is often advocated, in practice it rarely occurs. Finally, the emphasis of the programme on the evaluation of outcomes is a feature which is all too often absent from workplace health programmes.

Ireland - Social wellbeing in a rural area

Fruit of the Loom International Ltd. - McCarters Ireland Ltd. (FL) represent an almost unique industrial development in Ireland. Situated in the North-West of Ireland, it is located in what has for many years been one of Ireland's poorer and most underdeveloped regions. High levels of unemployment, emigration and a dependence on small scale agriculture characterise the area. McCarters originally was a family owned general clothing manufacturer employing in the region of 400 people in Buncrana (a small town with approximately 4000 inhabitants). When FL became involved with McCarters in 1987, a development plan for the new company projected growth to 2,650 jobs which would be sited at three locations within the region.

This growth has had considerable impact in the area. The local labour pool for women workers has proved insufficient, and inward migration from many parts of Ireland has occurred. A massive increase in cash flow through the local economy has resulted in a range of problems, in addition to the opportunities normally associated with economic development on this scale.

The climate within the company is best characterised as being open, positive and growth oriented and the developing occupational health services within the Company reflect this attitude.

The health action programme at FL has three principal elements. The first covers the provision of occupational health services. These include first aid, information provision, and counselling on a range of health issues, often focused on women's health issues. In addition, referral to medical services and outside agencies occurs. The company have instituted a "Wellness at Work" programme. This is a health promotion programme largely focusing on CHD which has been imported from the American parent company. Finally, "Health Awareness Evenings" have been run jointly with the local Health Authority. Further activities in the planning stage include an anti-smoking programme focusing an individual and environmental strategies, a healthy food policy in the company's canteens and the provision of cervical smear testing.

The second major element of the health action is an educational intervention. This action is not targeted at physical health issues but at the social integration and wellbeing of current and prospective employees. It focuses on providing information to young workers and prospective workers concerning the transition from school to work. Conducted both in the company's premises and in the local schools, it addresses issues concerning the management of income, the effects of employment on family life, planning for the future and the nature of working life.

Follow-up activities to this intervention are also provided, and these take the form of arranging for local banks, building societies, credit unions, tax advisers, insurance companies and travel agents to be available in the plant on a weekly basis for consultation by the workforce.

The third element provides an infrastructure as well as facilities aimed at promoting health. To date the company has provided a major part of the finance for the local area's only swimming pool. They have also contributed to the development of the town's water supply. In addition, the provision of other facilities, e.g. funding the building of a gymnasium and tennis courts are currently being examined.

Commentary

The Irish case study, which took place in the clothing industry in rural economically under-developed region, illustrates some of the problems which can occur with rapid industrialisation of such areas. Also, it provides a good example of the benefits of adopting a flexible approach to the problems which emerged. A particularly interesting feature of this case concerns the fact that the approach taken identified social issues and social wellbeing as being a major dimension of the problems which occurred. This led to a non-medical series of interventions which had the combined effect of preventing problems from arising in the first place and of providing the workforce with the skills to deal with problems which did arise.

This initiative, in common with the Greek initiative described below, also effectively addressed the issue of the wellbeing of the wider community. Underpining this approach is an overt model of health which explicitly recognises the social dimensions of wellbeing, and the role of appropriate life skills in promoting positive health. This case study also illustrates that effective health initiatives can take place in companies where there are no in-house medical (as distinct from nursing) services available.

The Netherlands - The "Healthier Work" project

This project was aided by the Ministry of Social Affairs and Employment and the Ministry of Welfare, Public Health and Cultural Affairs. It's aim is to improve the safety, health and wellbeing of workers and to reduce absenteeism and job losses due to disability and resignations. The project aimed to develop a methodology for application in other work situations. Beginning in 1987, a stimulus for the project was a report from the personnel department and the company doctor that serious problems (stress) were occurring in some work situations and that absenteeism was high.

Based on a study of 300 workers, it established that there were problems caused by stress at work which prevented workers from fulfilling their job requirements. The main causes of stress included materials not arriving on time, lack of support from bosses and workmates, being undervalued, low autonomy and a lack of meaningful work. It also emerged that coping skills were unsatisfactory. The main symptoms reported were low job satisfaction, emotional problems and physical ailments.

Based on the results of this study, management agreed to carry out the "Healthier Work" project. The aims of the project were to increase workers' feeling of wellbeing, to increase safety and reduce unsafe behaviour and occupational accidents, to enhance healthy behaviour and reduce risk-taking behaviour, illnesses and biomedical risk factors and to reduce absenteeism, job changing, disability and/or sickness costs.

The project focuses on three approaches to occupational stress. These are:

- Stress prevention: influencing the situation requirements and workers' own abilities;

- Stress management: training workers to manage tension between situation requirements and their abilities;

- Stress reduction: reducing stress already present in the individual worker.

An analysis of working conditions and work experiences will be made before interventions are made. Use will be made of the WEBA methodology, but also of other types of instrument and of open interviews. The findings from the survey will be translated into personal terms for each job in the company. An analysis will also be made of the educational status of the workers in relation to current and future job requirements - here both the technical and social aspects of the job will be taken into account. Finally, intervention modules will be developed, aimed at increasing wellbeing and promoting healthy living, such as social skills, stress management, leadership skills, physical exercise, smoking and healthy eating habits (including reducing alcohol consumption).

Employees will be maximally involved in the project. One way of achieving this will be to set up a trouble-shooting group. This group will make recommendations for change based on more general analyses of working conditions. These will then be tested in interviews with individual workers. The group will also analyse staff qualifications and make recommendations for adapting current job requirements to the existing levels, as well as recommending the creation of opportunities for further training. These recommendations will be tested in interviews with individual workers. The researchers envisage that the group will help create a supportive climate in the workplace.

A pilot project has been carried out at one of the company's plants. Its aim is to explore the feasibility and also factors which promote workers participation. Information meetings and interviews with workers are planned before

interventions take place. Efforts are also being made to involve workers as much as possible in introducing the interventions.

The evaluation plan is to take measurements before and after the intervention, and also to have two of the company's other plants act as controls. Measurements will be taken before implementation and again after one and two years. Variables to be measured include lifestyle (smoking, drinking, sleeping etc.), psychosomatic ailments, wellbeing at work and on working conditions, attitudes at work and contract conditions. Some physiological measures will also be collected including heart rate, blood pressure and breathing.

Recently two similar demonstration projects have started - one in a hospital and one in a building company. These projects have been funded by the Dutch Government and will be finished in 1993.

Commentary

The Dutch Healthier Work project again has the key feature of systematic approach to health problems in the workplace (in this case stress). The entire project is based upon the findings from a survey of stress and wellbeing amongst all of the employees of the company. The fact that it offers the opportunity for all workers to participate in the programme is also a significant feature of the project. A further significant feature of this project is the overt distinction between stress prevention, stress management and stress reduction. These distinctions mean that unlike many stress management projects in the workplace, it does not solely focus on improving the coping skills of individuals, but also looks at environmental factors as they may relate to stress. This case study also had the benefit of producing a methodology which could be more widely applied to other workplaces, this is evidenced by the interest of the Dutch government.

The UK - The Welsh Heart Programme

A number of issues influenced the selection of the workplace as a setting: more than 33% of the 2.8 million people in Wales are engaged in employment; workplaces in the service and retail sectors are visited by large numbers of the public; environmental factors can be partially controlled; a consultative framework often exists for workplace health and safety; and the corporate identity of the workplace could stimulate the development of shared positive attitudes to health.

The components of the Heartbeat-Wales Workplace programme include: anti smoking programmes and policies; an extension of OHS to cover, for example, counselling and health appraisal and Look After Yourself (LAY) classes. Menu labelling and LAY classes were begun in relation to nutrition, health canteen policies and greater dietary information. The exercise component included efforts to develop facilities and LAY classes. In addition, psychosocial health was addressed through self-help groups and LAY classes while training on cardiopulmonary resuscitation was also provided.

The development of the Workplace Programme involved an initial series of discussion meetings which aimed to bring together key actors relating to work and health for discussion of possible areas of collaboration. The support of the Welsh TUC, the British Institute of Management, and the local Health Authority was enlisted. Further meetings were held to sustain interest in workplace health promotion and to examine specific topic areas. Seminars were also held with other key actors such as caterers, ASH in Wales (Action on Smoking and Health) and relevant professional groups.

Heartbeat Wales produced a range of materials to support the Workplace Programme in collaboration with other key actors. These included information leaflets on workplace health promotion, the impact of CHD in Wales and how risks can be reduced and on healthier eating in the workplace. A Heartwise Menu Scheme was introduced to help caterers to promote healthier food choices for which support materials were provided and a recipe book was produced.

Several projects illustrate the collaboration between Heartbeat Wales and national organisations. These include the Heartbeat Award Scheme (designed for caterers) and the Make Health Your Business Award scheme for businesses. Applicable to both small and large firms, a company is required to submit a description of its health promotion strategy to qualify for the award. Heartbeat Wales has produced an information package giving guidelines to employers on how to achieve health promotion in the areas of smoking, nutrition, alcohol, exercise and stress. A mobile health screening service (Well Welsh) was also provided.

Finally, an educational project, "Look After Yourself" has been promoted as part of the programme. It is a national network designed to teach people the skills of sensible exercise, stress control and nutrition. Using local tutors, an advantage of the system is its flexibility, as tutors can respond to local needs and provide additional components according to demand. In some companies staff have been trained as LAY tutors and LAY classes are run in the workplace on company time.

Results from an ongoing assessment of the programme indicate that interest in health is growing. For example, 27% of companies have smoking policies, the majority of which had been introduced within the previous 4 years. 12% had a policy on healthy eating and 20% of companies had policies on alcohol or HIV/AIDS. Policies on stress (4%) and exercise (1%) were less widely reported.

More than 50% of companies offered health screening services, 38% of companies offered counselling services for staff, covering smoking, stress, dietary or alcohol problems. 31% of companies in the survey knew of the Well Welsh Service and 87% requested information about the service. 99% requested further information from Heartbeat Wales on workplace health promotion.

Successful aspects of the programme include smoking policy development and in the uptake of health screening. In addition, 70% of companies had at least one canteen on site and a quarter of these companies regularly ran healthy eating events and menu labelling schemes. Less successful areas include a low rate of provision of exercise facilities on site, while few companies operate group support classes other than for first aid training. Only 4 companies were running LAY classes at the time of the study. Furthermore, only 35% of companies in the survey operated on site occupational health services, and only half of these were served by a full-time nurse, and 3% by a full time doctor.

Commentary

The Welsh Heart Programme is a prime example of a major government health promotion initiative has been expanded from targeting the community at large to include workplaces as a specific area of activity. Even though it has concentrated on a single issue - heart disease - the principles of approach which involve environmental and behavioural modification are applicable to other health problems in the workplace.

The Welsh Heart Programme has instituted a number of activities in the workplace. These include educational activities in terms of provision of classes, the provision of educational materials and a number of specific projects. Projects have also been designed to improve the quality and choice of food available within works canteens and an awards programme for businesses and a mobile health screening service have also been developed. A key feature of the Welsh Heart Programme is its emphasis on evaluation activities. All of the activities which have taken place within the workplace have been systematically evaluated, the results of which are currently becoming available. This feature is all too often absent from many other workplace initiatives.

Greece - the evolution of in-company health services

This case study concerns the TITAN group of companies, founded in 1911, with a current work-force of 2043 employees. The TITAN group deals with every aspect of cement production ranging from extraction through to production, transport and sale of the finished product.

TITAN began to pay special attention to safety in 1973 following a fatal accident. As a result a permanent committee on accident prevention was set up in 1974. This committee advised that further safety measures be taken. From 1978, these committees included workers' representatives. Following the passing of new legislation in 1985, committees which existed previously were replaced by committees elected by workers. Today the safety service in the Athens office has two technical safety officers and an occupational medical officer. It directs and co-ordinates measures to improve working conditions, while leaving regional units to take further action considered necessary.

The OHS began in 1977. From 1980, occupational medical services were organized along the lines of those in other international companies and fully equipped sick bays were set up in each factory. Today, nine full/part-time occupational medical officers, together with support staff, are employed. This parallel development of accident prevention and occupational medicine has resulted in an overlap of the services provided.

Safety measures: Initially, efforts were limited to guarantee safe working conditions in factories--protective clothing was provided, fire-fighting groups were organized, and fire extinguishers and other protective systems were installed. Workers were trained in first aid. However, it was accepted that these preventive measures could not on their own significantly reduce occupational accidents, so parallel efforts were made to motivate and educate workers in health and safety. These measures included workers educational meetings; co-operation with staff and trade-union organizations and health and safety committees; staff training seminars; providing posters, slogans, stickers, and warning signs on accident prevention; publication and distribution of leaflets and pocket diaries containing safety advice; publication and distribution of a book on safety directives; competitions for producing posters and notices on accident prevention; safety questionnaires are distributed to staff and prizes awarded for correct answers; safety awards for factories with lowest rates of accidents; and prizes for staff for accident-free periods.

As a result of this campaign and the growing interest of workers health and safety issues, occupational accidents (OA) have fallen considerably. For example, the OA frequency index fell from 44.3 (1972) to 4.7 (1988). The reduction has reinforced workers confidence in the programme and increased their involvement.

Health services: Progress has also been made in improving workers health. In 1980 biannual chest X-rays were introduced, and in 1978 spirometric tests were introduced. Preventive health care was augmented with hearing tests in 1985; the goal for 1990 is to check cholesterol levels and heart disease. Attendance at medical examinations is voluntary, and uptake is high, e.g. 100% for spirometry, 80% for chest x-rays and 80% for cardiological tests. Results from regular examinations are analysed and correlated with in the work environment factors (dust, noise, etc.) to inform decisions on possible preventive measures.

Systematic health education efforts are also made. Since 1976 about 110 advice leaflets have been issued; 4000 copies of these have been circulated and distributed. In addition, posters are displayed throughout the factories and advertisements are inserted between the pages of the company diary.

Social Welfare Services: The TITAN group has had a welfare service for some years. It is concerned with health matters affecting workers and their families. In the field of preventive medicine, it helps organize PAP tests for female workers and workers' wives, immunisations (influenza, tetanus), organizes blood-donor sessions (providing a blood bank for workers and their families) and provides education to workers' wives twice yearly, on a range of social and medical issues.

TITAN's social and recreational programmes include: New Year's Day festivities, an annual outing, payments for disabled workers, children's camps, children's party with school prizes and Christmas presents, reception for students gaining places in university, and a carnival ball. Loans are provided for workers and an insurance scheme exists for staff and their families. Financial assistance is also provided to municipalities near the factories in the form of material and technical help and advice, particularly for projects of mutual benefit (e.g. roads, creches, hospitals, tree planting).

Commentary

The case study concerning Titan illustrates how health services within a company or group of companies can develop in a relatively short space of time. It also illustrates the evolution of workplace health services from an initial concern with safety in the workplace to a more general concern with health and wellbeing.

As far back as 1973 the safety activities in the workplace were considered to be at an inadequate level. However, following a fatal accident, these safety activities gradually increased over the years to the point where accident rates have fallen dramatically. A feature of the development of health and safety services within this company has been the fact that it now does not solely concentrate on safety, but also on health issues. Initially confining their health activities to medical screening, be it general or specific, they have also begun to become involved in health education programmes.

A further interesting feature of this Greek case study is that there are a number of welfare provisions provided by the company, which would appear to be an important adjunct to the company's health and safety policies and also to Greek industry at large. These welfare provisions are aimed interestingly enough, not just at the workers, but also at members of their family and would appear to have the potential to make a major contribution to workers wellbeing and also to their families wellbeing. Finally it should be noted that levels of participation by workers and their representatives in health and safety matters, appears to be high. It is possible that cultural factors in Greece, where relatively strong social provisions by companies appear to be common, have contributed to this high level of participation by the workers and their families.

The Netherlands - Improving VDU Work

This project on VDU work spearheads the Federation of Netherlands Trade Unions (FNV) action plan of working conditions. A stimulus for the project was research from the Ministry of social Affairs and Employment. This showed that headache, tension, muscular complaints and eye-strain occurred due to a combination of long periods of VDU work, heavy work pressure and a negative assessment of the working environment. In addition it indicated that there was a link between visual-ergonomic conditions and complaints of headache and eye-strain. Poor postural ergonomics was found to be linked to muscular complaints and headache.

Although many people had positive experiences of being able to work more efficiently with a VDU, many found VDU work too simple and wanted more variety in their work. The non-VDU jobs they had to do were generally more favourably assessed, in other words less stressful. One of the operators' main demands was for education material to be developed on the basis of this research.

The project adopts a two-pronged approach in providing information to the operators and also to their representatives in the companies (the unions and works councils). The aim of the project is to enable individual action and to offer a framework within which individual queries/complaints could be more effectively translated into some form of collective action.

The target group for the action are clerical workers (mainly women) working in the civil service whose work typically involve long periods of uninterrupted work at VDUs. The main source of information for the worker is a questionnaire which aids in the assessment of their own jobs and workstations. It addresses posture, readability, reflections, the work environment, tasks, and scope for organising and communication. In completing this questionnaire the respondent is assessing his/her workstation and job in terms of risks to safety, health and wellbeing. At the end of the questionnaire a 'risk profile' is set out which identifies areas of high risk according to the headings outlined above which focuses efforts to improve elements of the job and the workstation. In order to effect these improvements, suggestions are included in the information material supplied to the operator. For example, the worker might be directed to discuss the results with workmates and ask them to fill in a questionnaire, or to get in touch with their immediate boss or manager, or to make the contact with the company doctor. This advice is intended to broaden the range of individual options. In addition, expert support by telephone is available.

The documentation for union representatives/works councils also contains a list of VDU operators' potential needs. This list outlines possibilities for the purchase of new furniture or ergonomic testing of VDUs. From this list it is possible to generate requirements for workstations, work organisation and operator training. The package also includes directions on how to generate risk profiles for individual operators per type of job, per department or per organisation. It also contains suggestions for measures to improve VDU usage, such as listing the problems for those in charge, or starting a "VDU Black Book" which can be brought up in formal discussions with management, or meeting the personnel department to talk about possible solutions, or calling in the company doctor to underline the risks reported and increase pressure on management. Finally, the package contains a summary of the legal options open to unions and works councils.

The Union project commenced in October 1989. It has been publicised through the Union's own newsletters and in some of the popular women's magazines. So far, the results have been very encouraging. In a period of three months, almost 8,500 women have applied for the information package. Many complaints came to light, for example headaches and pains in the neck/shoulders, but also some regarding job content. Notably, only 20% of those who got in touch with the FNV were members of an associated union. In addition, approx. 300 works councils and union groups have requested material. Other job groups are also showing interest in the material. This material could also potentially be used in the health care services (physiotherapists and company medical services).

The second stage of the project involves CAD (Computer Aided Design) professionals. The questionnaire for clerical VDU operators has been upgraded and adapted for the new target groups. It will be introduced to architects, engineering consultants, computer companies etc.

Commentary

The principle distinguishing feature of the VDU work project is that it is emanated, not from company medical services or the companies themselves, or government agencies, but from trade unions and is workers' representatives. Again, like almost all of the other projects are outlined in this chapter, the basis for action has been the systematic analysis of what the nature of the problem. In this case the ergonomic and work organisation problems associated with the use of VDU's are the subject of enquiry.

Spain - Applying health promotion principles

Barcelona's Municipal-Transport Company (TMB) manages public transport (buses and under-ground trains) in Barcelona and its environs. It employs 4700 workers and has undergone restructuring in the past few years, a process that has affected the company medical service. The activities of this service were previously restricted to providing regular medical checks required by law. Following restructuring, a team of specialists responsible for medical services has radically changed the policy and focus of the service, and now offer integrated health care based on prevention and health promotion.

The OHS policy is now based on three key criteria: to aim health actions at the entire working population and not simply at high risk groups; to give preventive action priority over treatment; and to involve everyone in the company, particularly management and the trade unions, in health promotion. Health programmes now concentrate on two kinds of action: primary prevention based on health promotion programmes; and secondary prevention based on medical examinations.

Four programmes are currently established in TMB. The first concerns the prevention and monitoring of smoking, based on medical counselling for individuals concerned. The second focuses on the prevention and monitoring of alcohol abuse, based on an education campaign in the company magazine. This is based on an analysis of the extent of the problem and the selection of a method of detecting alcohol abuse, the aim of which aim is to facilitate medical counselling and the provision of care for those who may need it. The third programme is called "Tran-Sport", a programme to promote physical exercise the initial aim of which is to reduce the occurrence of back problems among bus drivers. The aim of the second phase of this programme is to encourage physical exercise and participation in sports activities on a regular basis, with a view to improving posture. The final programme is a stress-prevention programme aimed at management staff.

There have also been changes in policy on regular medical checks. These are now performed every two years rather than on an annual basis (only for employees who agree to this arrangement), a step which reduces the workload of the company medical service and helps to "demedicalize" prevention. Screening has also been introduced to diagnose preventable diseases such as breast cancer.

Assessment of results obtained from these changes is tentative as the programme has only operated for a short period. However, some of the hoped-for effects are being achieved, such as a 14% drop in the number of smokers. The other three programmes have not produced any tangible results as yet, though it would appear that the general level of awareness within the enterprise has been raised.

A number of problems associated with the change in health policy have been identified. To date there has not been a high level of collaboration with trade unions. This may in part be a consequence of the stigma attached to the company medical service, which, because of its previous policies, is seen as a service of the enterprise and not of and for the workers.

The hegemonic cultural model of health in Spain among workers, management and specialists is a major obstacle to any attempt to refocus the policies of a company medical service. Changing the system of medical examinations was a difficult task in that workers had to be persuaded to accept a change they saw as a management ploy to offer fewer services and to reduce social costs.

The involvement of trade unions and management in devising and implementing health policy was seen as the key factor of progress by the staff of the company medical service, whose aim is therefore to direct actions both at workers and at management personnel.

There is, however, no clear link between environmental prevention, aimed at monitoring and improving the work environment, and action concerning risk factors that are not solely or exclusively associated with the working environment. The development of such a link would help to increase the participation and involvement of the trade unions. It should be noted, however, that the workers' response was more positive than was that of the trade unions. Greater involvement of the company medical service in environmental issues may improve trade union support.

Commentary

In some senses the strength of this project is also its weakness in that it relies upon the provision of information, albeit targeted information to individual workers who then must act through their unions, their works councils or in cooperation with management to overcome the problems identified. Whereas the project empowers individual workers to enable them to perform appropriate analyses in an efficient way, it also potentially leaves them at risk as they may be perceived as being individuals acting alone. However, the uptake of the programme indicates that this may not be a significant problem, as well as signifying that a large demand for health and wellbeing programmes of this type exists, at least in the Netherlands.

Perhaps the most remarkable feature about the health promotion activities of Barcelona's municipal transport company is that the service offered by the occupational health department has consciously adopted principles of health promotion as opposed to more traditional treatment-oriented models. The giving of priority to preventive action over treatment, in addition to the reduction in frequency of medical screening activities provide concrete evidence of this commitment. This case study also illustrates the difficulties of achieving effective participation by the major actors within the workplace at least on an institutional level. The fact that medical services were being reorganised gave rise to certain suspicion among trade unions. Despite this suspicion it appears that the difficulties associated with this transition have been at least to some degree overcome.

Some More Limited Examples

Sweden - A Practical Example of Improving Wellbeing at Work

The Scandinavian model of work organisation and participation has long been considered to offer potential for improved wellbeing at work. One of the major examples from Sweden concerns the LO (a Trade Union Federation) project where a nationwide workplace health programme has been instituted. From the many examples of work organisation and participation which are available an interesting case which illustrates how this concept can be applied even in very small enterprises is provided by Haglund et Al (1991).

This case study describes changes in work organisation undertaken in a small clothing manufacturers in Sweden, employing 17 seamstresses. They have moved from a traditional production line form of assembly to a team working approach where all seamstresses are capable of undertaking all tasks in relation to the making of a garment. This change was inspired by a situation where the company's financial viability was threatened. As a response team-work was introduced. However, a major consideration in introducing this alternative form of work organisation was the improvement of wellbeing among the workforce. A crucial aspect of the success of this new organisation of work was the high level of involvement and participation of the workers in decisions about how work must be undertaken and about the general direction of the company.

Whereas traditional assembly line work was characterised by time pressure, monotony, repetition and mistakes, the new work organisation is far less monotonous and repetitive though time pressure still exists. The company has become more flexible in its approach to its business and is capable of turning round orders in a shorter space of time thereby maintaining time pressure as a feature of the work environment.

The effects of this change in work organisation were quite dramatic. Levels of sickness related absenteeism and turnover reduced considerably and morale within the workforce improved correspondingly. The traditional illnesses associated with this kind of work (e.g. musculo-skeletal pain) have been reduced, though there have been some re-occurrences of old injuries there has not been an emergence of new complaints of this type. In addition, there is a rehabilitation programme for those who have long-standing work related illnesses, and work site exercises to help prevent special injuries from occurring in the first place.

Commentary

Although this is a small scale case study it provides a very clear illustration of not alone the benefits to be gained from undertaking changes in work organisation, but also the integration of health and wellbeing issues into the change process of moving from traditional forms of work organisation to more advanced forms. The strong emphasis placed on high levels of participation by the workers is also a key feature of this case. It is interesting to note that one of the reasons why participation succeeded so well in this instance was that a new manager had been brought in to the company who had previous experience of working in a participative manner.

There are other case studies which illustrate some more limited aspects of good practice of workplace health actions and can be contrasted with more traditional approaches to, not alone occupational health, but also workplace health action. Examples of some of the more limited approaches to health action in the workplace can be found in Marshall and Cooper (1981).

For example, Sworder (1981), describes a research project and training programme which is undertaken by ICI in relation to stress in the job. In this programme, a survey was conducted and a range of typical management stresses were identified. In response to these stresses a training intervention was planned and implemented during normal management training courses. The interventions sole aim was to improve management coping skills, and it is notable that stress not was found to be "high key issue". Despite the diagnosis of a range of sources of occupational stress related to organisational structure and culture, no attempt was made to address the environmental correlates of stress in the workplace. This can be contrasted to the approach of the Healthier Work project in the Netherlands described above.

Another example of a limited approach to health promotion in the workplace was presented by Colacino and Cohen (1981), where they describe a 'total health and fitness' programme as a response to the issue of health in the workplace. A number of individual programmes were offered under this initiative, these include an employees rehabilitation and recreation programme. Whereas the authors recognise the difficulties with evaluating in particular, the long term impacts of approaches such as this, it is noticeable that this programme, which claims to be comprehensive, is directed solely at the alteration of individual behaviour.

There is a good deal of interest in this kind of programme in US companies, which present a different corporate culture to many European companies. The philosophy of having healthy and fit employees as a contribution to corporate effectiveness would appear to have taken root to a much greater degree in the US than it has in Europe. Many examples of this approach can be found in the literature (see Sloan et Al, 1987, for further examples). However, there is an increasing recognition of the limitations of these kinds of programmes (e.g. Fielding, 1990), and attempts are now more likely to be made which seek to integrate both environmental and individual interventions.

2.3 So what are we talking about?

In essence workplace health promotion includes much traditional activity in the workplace but it is a larger concept than is usually evident in traditional activities. It does not particularly seek to differentiate between sources of threat to health which arise in the workplace and outside, but seeks to influence the health of the individual in a positive way regardless of the source of the threat to health.

The examples of above provide a wide ranging selection of some approaches to workplace health. In many ways these are examples of very good practice in that some of them (for example the Italian and UK case studies) incorporate large scale activities across many enterprises using a genuine variety of methods and involving multiple disciplines in the implementation of the health action. On the other hand some of the smaller case studies for example the Swedish one, illustrate what can be done on a small scale enterprise with an imaginative approach.

It should also be noted from the above examples, that in many cases the good practice consists of recognising that there is a problem, and formulating a plan to overcome the problem. Case studies from Ireland and Portugal took place in economies and contexts which are not particularly associated with sophisticated approaches to health issues. Again these illustrate what is possible even in the most unpromising of circumstances and also it depends in many cases an innovative approach to health in the workplace consists of implementing actions for the first time. By contrast the German and Dutch case studies show the virtues of planned approaches in more developed settings.

In some ways none of these case studies live up to the five goals or principles of approach to health promotion outlined earlier in the chapter. However, an examination of them all shows that they were all informed by the spirit of these principles (even if they were not consciously applying them) in many of the actions which they undertook. In each of these examples three or four of these principles of approach have been applied successfully.

3. The Playing Field - The Official Story.

This chapter outlines the current legislative provisions operating within the European Community and particularly within the eight member states which took part in the study. In addition there is an account of the attitudes or orientations of the major actors on the scene of workplace health. This material is drawn from the interviews which were held in the first phase of the study. This chapter emphasises the major constraints and opportunities for workplace health action which are provided by both legislation and the orientations of the social partners.

3.1 Introduction

There are large differences in the legislative background relating to workplace health in the eight countries examined. At time of writing there is no distinctive European legislation which is expressed in common provisions across the eight member states, rather a collection of laws which reflect different philosophies and approaches, and different levels of economic and industrial development. The effects of historical changes in political structures in Europe also play a role in the shaping of legislation, where, for example, much of Spanish legislation emanates from Franco's time, while in Ireland the residual influence of British legislation is still felt.

The completion of the internal market, and particularly the implementation of the provisions of the Maastricht treaty at the end of 1992 will change that situation at least to some extent, in so far as the relevant EC Directives must be implemented into national legislation. While this process is well underway, and provides an example of the influence of EC level legislation, it is still possible that there will be national differences remaining, as the distinctive political cultures will remain, and indeed, the precise details of implementation in each country are likely to differ.

Despite these differences between the countries, the various national legislation instruments, even as they stand, are implicitly if not explicitly supportive of workplace health actions of the kind of interest to this research. At worst, legislation does not address the possibilities for health promotion in the workplace, while at best it provides a framework which legitimates such workplace actions. However, in no case does legislation direct activity in this area.

It should also be recognised that in many countries workplace health issues are also regulated by non-legislative instruments such as ministerial regulations and in some countries (e.g. Italy, Germany) by collective agreements between the social partners. However a review of these instruments is beyond the scope of this report.

3.2   Legislative developments

3.2.1 International developments

The signing of the Maastricht Treaty for European Union (at time of writing it is currently undergoing the ratification process in all 12 member states, and in very recent times the Danish people have voted marginally against the adoption of the treaty) has for the first time given the European Community this role in the area of public health. It particularly recognises that the health of the people of Europe depends not only on the provision of health care and health services in the various countries, but also recognises the need to promote access to health information and education and the need to consider health protection as part of community policies.

Of course some of the major implications of the ratification of the Maastricht Treaty remain to be developed. Quite how the provisions in the public health sphere will be applied, and precisely what their implications will be for workplace health promotion is not yet clear. However it seems possible that the emphasis on information and education could alter the climate for workplace health promotion in a positive way.

Related to the Maastricht Treaty has been the signing of the Social Charter. The development of the internal market EC by the end of 1992 has encouraged the development of counterbalancing measures which seek to guarantee workers rights. In this regard the Social Charter [the Community Charter of the Fundamental Social Rights of Workers], which has been signed by 11 of the 12 Member States and the Framework and related Directives on safety and health at work have an important role to play in setting the context for the development of workplace health action throughout Europe.

The Social Charter which was agreed by 11 of the 12 EC governments (the UK being the exception) at Maastricht contains a number of provisions which have direct and indirect relevance to the development of workplace health action. Successive drafts of the charter have reflected pressure from some governments for its provisions to be weakened. Despite this pressure the Charter still promotes the improvement of living and working conditions and constitutes a considerable advance in this regard over the existing situation in many of the member states. Among its provisions are measures to promote the improvement of conditions of work; work contracts; social benefits; freedom of association; information, consultation and participation; and health protection and safety at the workplace. Though the Social Charter was not formally part of the Treaty signed at Maastricht, (it was the subject of a formal agreement outside of the Treaty), the provisions of the Charter must ultimately be enacted into regulation and legislation by the Member States. These enactments of the Charter will then function, *inter alia*, to support the development of health protection in the workplace and should provide a basic framework in which and increased and broader level of workplace health activity can take place.

ILO Convention 161

The ILO Convention 161 (1985) on Occupational Health Services (OHS) has also played a role in setting the scene for generating more workplace health actions. This convention emphasises the preventive role of OHS, and also includes the protection of mental health as one of the functions of OHS. Other important themes of Convention 161 include:

- the need for consultation with representative bodies in the monitoring and evaluation of OHS policy and practice;

- the right of workers representatives to participate in OHS practice;

- the need for an appropriate multidisciplinary approach to the provision of OHS;

- the need for the independence of OHS; and

- the right of workers to information about workplace health hazards.

The Framework and other Directives

ILO Convention 161 has also influenced the development of policy by the EC. The adoption by the Council of Ministers in June 1989 of the Framework Directive on Health and Safety was influenced by Convention 161, and marks the spearhead of a number of subsequent Directives which deal with a range of workplace health and safety issues. This Framework Directive adopts lays down strict obligations on employers and workers, provides for:

- the establishment and maintenance of prevention, protection and emergency services;

- comprehensive information and training in the area of health and safety; and

- full consultation and participation rights to workers on matters affecting workplace health and safety.

This and other Directives will also help promote workplace health actions as they will oblige governments to enact legislation which in many cases exceeds current provisions in these areas. As an example, the display screen draft Directive provides for the application of the principles of both hardware and software ergonomics, a position which represents a considerable advance on current legislative provisions in many countries.

A further Directive, on the "Introduction of measures to encourage improvements in the safety and health of workers at work" is explored in some detail below to provide an illustration of the overall approach adopted. An important aspect of this Directive is that it does not allow for the "rounding down" of provisions for safety and health between countries. Other important aspects of this Directive include provisions to ensure that:

- Workers and/or their representatives must be informed of risks to safety and health and have the right to participate in ensuring that necessary protective measures are taken;

- Information, dialogue and balanced participation on health and safety at work must be developed between employers and workers;

- The improvement of workers' safety, hygiene and health at work must not be subordinated to solely economic considerations; and

- Employers are obliged to keep informed of the latest advances in technology and scientific findings concerning workplace design and to inform workers of these so that a better level of protection of workers safety and health can be guaranteed.

The three developments related to the completion of the internal market (The Framework Directive, the Social Charter and the Maastricht Treaty), though not directly influencing the prospects for workplace health promotion none the less help put the issue of health promotion on the agenda more widely and should eventually have some influence on attitudes at workplace level such that health promotion will become more likely.

3.2.2 Legislation in Germany

Two types of legal provision govern occupational health and safety in Germany:

- National legislation (Acts and Ordinances)
- Accident prevention regulations issued by the Industrial Mutual Accident Insurance Associations (Berufsgenossenschaften)

These two sets of regulations complement one another. The state authorities and the Insurance Associations are interested in the prevention of occupational accidents and disease on human, social, political and economic grounds. These regulations are monitored by inspectors of the federal states and of the Insurance Associations.

The German Occupational Health and Safety Act is the most important framework legislative provision. Under this Act the employer is responsible for workplace health and safety. The Act is basically a framework act - it specifies the employers basic responsibilities, the duties of plant physicians and safety officers and the organisation of occupational health and safety within the plant. While it provides for the conditions under which company medical services should be set up, the Act does not prescribe the kinds of activities they should engage in. This lack of prescription allows for almost any activity to take place, but does not require the company medical services to take any _specific_ measures.

Apart from the government agencies, three other agencies involved in occupational health are covered by German law: the Trade Cooperative Associations; Trade Supervisory Offices; and negotiations between the industrial partners.

Trade Cooperative Associations in Germany act as industrial accident compensation funds, and membership is mandatory for all employers, as are the regulations they set up to prevent occupational accidents. Controlled jointly by employers and trade unions, they are largely concerned with issuing regulations on technical matters, hazardous substances etc. Despite their experience in the field, the Trade Associations have been reluctant to become involved in innovative workplace health actions.

Trade supervisory offices, managed by the Laender, issue a wide set of binding rules, prescriptions and instruments as the Trade Cooperative Associations do. However, they are more concerned with industrial hygiene problems than matters more closely related to health. In practice, their activities overlap with those of the Trade Cooperative Associations. In addition, their responsibility for both giving advice and for enforcing regulations leads to potential conflicts of interest.

Negotiations between the industrial partners can take place on almost any health and safety issue. It can occur at two levels - plant level through works councils, and through collective bargaining at supra-plant level. The agreements reached in these negotiations acquire legal status.

In conclusion, there are legal options to set up innovative workplace actions, but there is no specific incentive for further more directive legislation to be passed. In practice, the implementation of new approaches depends on the goodwill of employers and/or the bargaining power of workers. However, the German government has recently undertaken a series of initiatives to improve the implementation of health and safety regulations, and health promotion programmes are seen as one strategy for modern occupational health and safety approaches. Finally, a comprehensive law on work conditions and prevention is currently being discussed.

3.2.3 Legislation in Greece

Labour law provisions in Greece go back as far as 1911. This has been supplemented by numerous Executive Decrees which specify in detail employers obligations to implement measures in certain occupations and working conditions, e.g. industry and crafts, construction sites. Special Regulations have also been put into place which govern health and safety in various industrial sectors, as have a range of laws protecting workers from specific environmental hazards, e.g. benzene, radiation.

Contemporary legislative developments have been informed by a number of more progressive criteria. These include the prioritisation of prevention, addressing health as well as safety, the integration of occupational health policy into general health policy, a multi-perspective approach to health and safety (e.g. work organisation, production of hazardous substances), and provision for participation by the social partners and specialists.

These criteria have informed the enactment of a 1985 law which main provisions include obliging companies with more than 150 employees to provide a safety and medical service, allowing workers to elect safety and health committees, establishing criteria for work organisation and location, and the prevention of hazards and physical, biological and chemical threats to health. To date the implementation of the Act appears to have worked well, with high numbers of affected companies providing medical and safety services. Progress on the issue of workers health and safety committees has been slower.

Bilateral agreements between employers and unions may also address safety and health, but this possibility has been little used to date.

Other provisions to promote health come from legislation in the health area. In this regard, health centres in Greece carry an occupational medicine brief. Legislative provision is also made for the Social Insurance Foundation to promote occupational health.

3.2.4 Legislation in Italy

The Italian Constitution defines the general rights and responsibilities in the area of health which include rights to health, work, the duty to protect health and the right for individual health not to be compromised by private economic initiative. The Civil Code obliges employers to protect the physical integrity of workers and to provide compensation for avoidable injury. In addition, the Criminal Code also contains a number of provisions relevant to occupational health and safety. (See Wynne, 1990, and particularly Garzi and Tonelli, 1990 for a vivid illustration of the practice of these provisions).

For the past number of years there have been two general laws concerning accident prevention and workplace hygiene (DPR 547 of 27.4.55 and DPR 303 of 19.3.56). These laws have a general field of application with the exception of certain sectors covered by specific provisions (mines, quarries and transport).

In 1970, after strong representations by trade unions, the Statute of Workers Rights was passed. This Act states, *inter alia*, that workers representatives have the right to monitor the implementation of legislation on health and safety and to promote research, and the elaboration of suitable measures to protect their health.

In 1978, general reform of the organisation of public health took place under the Health Reform Act (1978). It was based on the principles of integrated health prevention, planning and decentralisation. In addition, it transferred responsibility for health and safety to the Ministry of Health.

Three levels of intervention were outlined. Firstly, at local level through the establishment of Local Health Units and Health and Safety Departments which have the tasks of intervention, control, surveying, training etc. Secondly, at the regional level, which have responsibilities for planning and finally at national level (the Ministry of Health) which has the tasks of general planning and the development of legislation, in addition to supporting the Instituto Superiore de Sanita and the ISPESL (Safety Institute), which has a tripartite structure). The responsibility for workplace health and safety residing with the health ministry is unusual in the European context.

The innovative character of the Health Reform Act resides in Article 20, which defines the risk prevention activities in which Local Health Units must engage. These activities refer not only to hazards which can be linked to production processes, but also to "non-specific pathologies" which may lead to serious illness. Consequently, prevention and promotion activities can have a broader focus.

On March 27th 1992 an important law (no. 257) has been approved after a big trade union campaign. The legislation prohibits the extraction, manipulation, production, import and export of asbestos and products containing asbestos. This comprehensive law makes provisions for environmental decontamination, and for training and technical support. It also makes provision for the linking of environmental and workplace action.

However, problems of non-compliance and delay have characterised the implementation of the Health Reform Act. The co-ordination of all health and safety legislation has not yet taken place. A recent Parliamentary Commission and the impact of EC legislation have provided fresh impetus to its implementation. Recently, (1991) the Italian Parliament passed the necessary legislation to implement a number of EC Directives, though not yet the Framework Directive.

An integrative and innovative mechanism for promoting workplace health has been played by collective agreements between employers and trade unions. In recent times these agreements have defined a new participative approach to health, safety and environmental issues. New bilateral commissions have been established at both plant and national level. These commissions are helping to overcome the prevailing conflictual approach between management and unions, and offer a new chance to simultaneously improve workers health and company image.

3.2.5 Legislation in Ireland

Historically, Irish occupational health and safety law originated from legislation which pre-dated the formation of the State. New legislation was passed in 1955 and 1980 (the Safety in Industry Acts) which went some way towards modernising health and safety legislation. These Acts were largely traditional in focus and were limited in the range of workplaces which they covered.

Following the recommendations of the Barrington Commission of Enquiry (1983), the Safety, Health and Welfare at Work Act (1989) was enacted. Barrington's criticisms of existing legislation and practice fell into three general categories - the excessive legalism of the system, the limited scope of existing legislative provisions, and defects in the implementation of the system.

This Act has two main purposes - to streamline existing legislative provisions and to improve the enforcement process through setting up a National Authority for Occupational Safety and Health (the Authority). To the extent that it fulfils these purposes, the Act answers all three of Barrington's criticisms - it unifies much existing legislation, it extends the range of application of previous legislation to all workplaces and it provides a framework for the improvement of enforcement.

The 1989 Act provides a useful framework for the promotion of the kinds of workplace health activity of interest in this research. For example, the general duties of employers to employees refer inter alia to duties to design, provide and maintain a place of work that is safe and without risk to health, to provide systems of work that are planned, organised, performed and maintained so as to be safe and without risk to health and to provide such information, instruction, training and supervision as is necessary to ensure the safety and health at work of employees.

A particularly important aspect of the 1989 Act concerns the obligation on every employer to produce 'Safety Statements' which must detail the provisions made by the employer to secure health, safety and welfare at work. The directors of a company must supply an annual report on the extent to which the measures outlined in the Safety Statement have been carried out.

A further innovative aspect of this legislation is that it established a National Authority for Occupational Safety and Health. The Authority has, *inter alia*, a range of powers to enforce health and safety policy which did not exist heretofore.

3.2.6 Legislation in the Netherlands

New legislation was passed in the Netherlands in 1980 (the Working Conditions Act - WCA). The basis for this legislation was that employers and employees would be responsible for safety, health and wellbeing in the workplace. An innovative aspect of this legislation was that it encompassed the concept of wellbeing for the first time in Dutch law. (This provision appears to be unique amongst EC member states). In addition, the idea of responsibility was apportioned for the first time to both employers and employees. The old 1934 legislation (the Safety Act) exclusively placed responsibility on the employer. However in placing this dual responsibility, the responsibility is one of collaboration and participation.

The definition of the phrase wellbeing is much more limited than the one in vernacular use. Specifically, it refers to: suiting the work situation and the job to the capabilities of the worker; allowing work to be done as far as is possible at the discretion of the worker; allowing the worker to communicate with workmates while at work; and providing the worker with feedback regarding the purpose and results of the work.

The employer is charged with ensuring the greatest possible safety, the best possible protection of workers health and the promotion of wellbeing when organising work, setting up the workplace and determining production and working methods. In concrete terms, this means that safety and health risks should be reduced or prevented at source, the use of tools, equipment and substances hazardous to health and safety must be avoided, workstations, work methods and aids for working must be ergonomically suited to the worker and the content and allocation of jobs must take account of the personal characteristics of the employee

All of these obligations are subject to the condition that it is reasonably practicable to carry them out. These provisions oblige the employer to have a clear policy on safety, health and wellbeing which should be integrated into the everyday operations of the organisation.

Industrial organisations employing more than 500 people are obliged to provide a company medical service. In practice these obligations are often exceeded. The purpose of the medical service is to protect and promote workers health as it is affected by occupational conditions. This is to be achieved through making recommendations on aspects of company and working practice, conducting examinations, co-operating in accident prevention, promoting measures to reduce absenteeism, keeping informed about physical working conditions and analysing and reporting working conditions linked to occupational disease.

In addition to these provisions the WCA allows for safety officers and committees which operate in parallel but are linked to the medical service. Joint annual programmes and reports must be drawn up by these services.

Allowance was made in passing the WCA for its implementation in stages in order to promote the best possible way for employers and workers to arrive at improved working conditions. The aim was to have this process completed by 1990, by when the Act should have covered almost every Dutch employee. In practice, however, the implementation of the Act has been sketchy, as both management and workers appear to be relatively ignorant of its detail. Research on its effectiveness is inconclusive, but it appears to have had a positive if small effect on the practice of workplace health action.

3.2.7 Legislation in Portugal

Until the signing of the single European Act in 1986 much of Portugal's legislation in the area of health and safety at work dated back as far as 1958 (this coincided with the first wave of modernisation of Portuguese economy). This legislation enshrines a narrow view of the relationship between work and health. Labour law in Portugal has not been recognised as a distinct branch in law, but is part of the civil code.

The constitutional provisions in the 1989 constitution adopt a broader view of workers rights to health. In addition, the new health law (1990) identifies the priorities of health policies as involving health promotion and the prevention of disease, and also equity in the delivery of health care.

Portugal also has a mechanism between the social partners (Permanent Council for Social Cooperation) which has the capacity to make agreements concerning health and safety at work. The new Law on Safety and Health in the Workplace (1991) is a good example of the consensus on the need to update Portuguese legislation concerning these specific matters. It's framework of reference is ILO Convention 155 (but not 161) and EC Directive 89/391. Emphasis is given to safety rather than health and to the prevention of diseases and accidents rather than to health promotion.

Accordingly, with this new Law, the workers and their representatives have limited rights to participation in health actions in the workplace. Basically the right to be informed (e.g. about the introduction of new technology) and consulted (e.g. training programmes, and the right to elect health and safety representatives (the maximum is seven for organisations employing more than 1500 workers). However, there are no legal provisions for the setting up of health and safety committees in Portugal. Where these exist, they have been set up largely at the initiative of employers through the collective bargaining process.

3.2.8 Legislation in Spain

Spanish law is rather contradictory, since much legislation which pre-dates democracy still exists. In addition, major changes are required to bring legislation into line with EC Directives.

As a background, the Spanish Constitution recognises the right to health protection and that public authorities are responsible for monitoring safety and health at work. In effect this means that occupational health is both a public duty and a contractual right.

In general Spanish legislation is seen as obsolete. For example, the General Ordinance on Health and Safety at Work (1971) is a technical text which reflects no political objectives, and was passed at a time when trade union activity was clandestine. The Workers Statute (1980) provides workers with rights to physical wellbeing and an effective health and safety policy, while also guaranteeing rights to training and information and involvement in monitoring and inspection activities. Workers are also obliged to partake in training and to observe health and safety measures. Other legislation exists in relation to specific hazardous substances.

Other legislative provisions include obligations on employers to take out insurance for occupational health and safety and obligations for enterprises employing more than 100 workers to provide a company medical service, and the General Social Security Act (1974) named health and safety at work as a function of the social security system. At present, a new Health and Safety at Work Bill is ready to enter parliament.

Legislation in the area of health also plays a role. The General health Act (1986) mentions occupational health as a broad ranging concept, with responsibility for its promotion invested in regional health services who are to collaborate with labour authorities. However, this Act did no more than set out political objectives, with the result that there are regional differences in the services provided.

In general, with the exception of the General Health Act, health promotion is not mentioned in Spanish legislation. Furthermore the institutional bodies involved in occupational health are under-resourced. However, the Spanish legislation does not prevent innovative action, and its establishment is perfectly feasible even against this discouraging legislative background.

3.2.9 Legislation in the United Kingdom

Health and safety legislation in the UK dates back nearly 200 years. Numerous separate pieces of legislation range from the Health and Morals of Apprentices Act of 1802 (which was concerned with fixed hours of work, washing of walls, ventilation and voluntary inspection) to the Control of Substances Hazardous to Health Act (COSHH) of 1988 (which is concerned with hazardous substances and their monitoring).

Important pieces of legislation include the Factory Act of 1833 (which made provisions for factory inspectors), the Health and Safety at Work Act of 1974 (which set up the Health and Safety Executive (HSE) and Commission, and placed complementary responsibility for enforcement on the state and employers), and the Safety Representatives and Safety Committee Regulations of 1977 (which gave a legal basis for appointing trade union safety representatives). The main focus of this legislation has historically been on safety, and placing responsibilities on employers and/or employees to ensure that the working environment does not pose a hazard to the health of either workers or the public.

In recent years, the provisions of this legislation has been interpreted more broadly, and now encompasses broader health issues, particularly with regard to smoking. For example, a ruling in 1968 found that employers must take reasonable steps to control hazards for which they can be reasonably assumed to have knowledge of. In practice, this has meant that measures to combat smoking fall within this ruling. The provisions of the Public Health Act of 1933 (providing sufficient ventilation and keeping the atmosphere free from "noxious effluvia") have also been interpreted in this way.

The Health and Safety at Work Act of 1974 superseded many older and fragmented pieces of legislation, and provides a more comprehensive framework for health and safety in the workplace. It is similar to other framework pieces of legislation, as it places duties on employers to keep the workplace free from hazards to health and safety so far as is reasonably practicable. Its provisions have also been used in support of anti-smoking policies.

Under this Act, a number of key agencies have been identified which have a role to play in the maintenance of safety and health. Furthermore recent HSE reports (HSE, 1978, 1991) have identified the workplace as an arena for more general health promotion, not exclusively tied to workplace hazards. As a result, occupational health services are increasingly addressing these broader health issues.

Five principal agencies (which have legal standing) are involved in health actions in the workplace in the UK:

- The Health and Safety Commission, which has responsibility for the development and enforcement of health and safety law;

- The Health and Safety Executive, which enforces and implements the law;

- The Employment Medical Advisory Service, which is the medical arm of the HSE;

- Environmental Health Officers, who are an inspectorate employed by local authorities; and

- Occupational Health Services, who operate in individual companies. Employers have no obligation to provide these services, and their brief is usually narrow.

Future developments in UK law are likely to increase the scope for innovative action to be taken. In particular, the draft safety framework proposed under the Single European Act will advance the possibilities for such actions. There is also a White Paper on the Health of the Nation under parliamentary consideration. However, the 1974 Act has not been widely embraced, largely due to a lack of resources for the HSE which has led to a lack of monitoring activity. Accordingly, there is much room for the improvement of practice.

3.2.10 The main themes of legislation in the eight countries

Despite the fact that the eight countries have had widely differing recent histories (legislation dating from other jurisdictions, legislation dating from previous political systems), there seems to be a remarkable convergence (though with some notable exceptions) in the type of legislation currently on the statute books. This convergence does not seem to be wholly due to the influence of the EC, as many countries reported a slow implementation of EC directives.

Legislation in all eight countries with the exception of Spain can be characterised as framework legislation, i.e. legislation which sets down general principles for operation rather than listing specific provisions for specific situations. These framework Acts usually have a range of specific Acts subsumed under their provisions. The advantage of this kind of legislation resides in its capacity for application to all or most workplaces, and to establish common structures and principles for OHS. All of the countries previously operated without such framework legislation, which resulted in a piecemeal approach to health and safety issues, and a commitment to often outdated standards.

The prime locus of responsibility for OHS resides with employers in most of the countries. Typically, they have a range of responsibilities, which usually include the provision of safe workplaces, a duty to consult/involve employees in safety and health measures, and a duty to monitor their workplaces for hazards. Employees typically have rights to participate in health and safety measures.

Most of the national respondents were of the opinion that the legislative position in the eight countries (again with exception of Spain) was reasonably adequate to cover the possibilities for innovative health actions. However, the practice of health and safety in the workplace was not as widespread as the provisions of the legislation, either in terms of traditional health and safety actions, or more particularly in terms of innovative health actions. Part of the reason for this may reside in the almost uniformly low level of policing of legislation. Lack of resources, conflicts of interest between relevant agencies, and unclear definitions of responsibilities have all contributed to this lack of implementation. Against this background it may be more appropriate for legislation to provide incentives for health promotion to take place, rather than to act as a regulator or controller of such activity.

A number of other common legislative themes emerged from the reports from the eight countries:

- Relatively low levels of activity in occupational medical services. In Germany and Ireland in particular few obligations were placed on these services. In addition, the numbers of people involved in these services was very low in some countries, notably in Ireland and Greece.

- The question of interpretation of legislation was seen as being crucial. Legislation itself was seen as being no more than enabling, but the interpretation of legislation, either by key agencies or by the courts determined whether the spirit of the legislation emerged into practice. In almost all countries, broader interpretations of the legislation were possible.

- Relevant legislation for health in the workplace in all cases came from Health and Labour ministries. In all countries, the practical responsibility for implementation and monitoring resided with Labour ministries or its agents with the exception of Italy where the Health ministry and its agents were responsible.

- The legislative situation in countries will change in the near future as a result of EC agreements.

3.3 The attitudes of key actors

Interviews were conducted in each of the eight countries to assess the levels of attitudinal and practical support for workplace health actions with a range of key actors. These focused on such issues as the levels of awareness of different authorities, the level of co-operation and conflict between potential providers and levels of support for workplace health actions by key actors. It should be noted that these interviews were carried during 1990, and that some changes in the positions of the various actors may have taken place since then. Where evidence of such changes is available, the accounts of the interviews have been altered accordingly.

3.3.1 Who was interviewed ?

Table 3.1 provides a summary of the type and number of interviews held with a range of key actors in the eight countries. The studies assigned different amounts of effort to the interviews which were carried out, with no less than 32 (out of a total of 87) interviews being carried out in the Spanish study, while as few as three were carried out in others. Nonetheless, the interviews appeared to be judiciously selected, as they provide a wide ranging insight into the potential for innovative workplace health action in the various countries.

Table 3.1 Summary of the interviews conducted

Country	Interviewees	Number	Type of Body
Germany	Insurance	3	Sickness funds (AOU, BdB, TUU)
	Trade unions	1	Congress of trade unions (DGB)
	Employers	1	Employers organisation (BDA)

Table 3.1 continued. Summary of the interviews conducted

Country	Interviewees	Number	Type of Body
Greece	Government	2	Ministries for Labour, Health, and Welfare and Social Insurance
	Employers	1	Confederation of Greek Industry
	Unions	2	General Confederation of Greek Workers Syndicate of Workers and Employees of Athens
	Professional Associations	2	Society for Occupational Health Society for Ergonomy
	Others	2	Society against smoking, Athens School of Hygiene
Italy	Magistrates	2	Magistrates experienced in H&S
	H&S Tuscany	1	H&S prevention service in Tuscany
	LHU/SNOP	1	Occupational physician in Local Health Unit and member of National Association of Prevention Workers
	Trade Unions	2	Representatives from national unions
	State Company	1	State company medical service
	CNA	1	National Confederation of Artisans
Ireland	FIE	1	Employer representatives
	ICTU	1	Trade unions organisation
	HSA	2	Authority for occupational safety and health
	HPU	1	Health promotion unit of Ministry for Health
	FOM	1	Faculty of Occupational Medicine
	Others	2	Voluntary bodies dealing with specific diseases
Netherlands	Government	2	Ministries for Social Affairs and Employment, and Welfare, Public Health and Cultural Affairs
	Employers	1	Employer Representatives
	Trade Unions	1	Trade unions organisation
Portugal	Government	6	Ministries of Employment and Social Security, and Health
	Industrial associations	4	A range of trade unions and employers Associations
	Health Professionals	1	Portuguese Society for Occupational Medicine

Table 3.1 continued. Summary of the interviews conducted

Country	Interviewees	Number	Type of Body
Spain	Trade Unions	8	Trade union organisations in different regions
	OH services	5	Company medical services
	CSH	5	Health and safety centres
	INSHT	4	National Institute for Health and Safety at Work
	Municipal OH	3	Municipal Occupational Health Centres
	Regional Govt.	2	Departments of health of autonomous regions
	Insurance	2	Employers mutual insurance associations
	Others	3	
United Kingdom	Employers	2	Confederation of British Industry
	TUC	3	Trades Union Congress
	HSE	2	Health and Safety Executive
	EHS	1	Local Authority Environmental Health Services
	IOH	1	Institute of Occupational Health
	GM	1	Gardner Merchant - Contract catering company

3.3.2 Attitudes and actions of key actors in Germany

In Germany, sickness funds have been in the vanguard of actions in the field of research and the implementation of general and specific workplace health promotion programmes. These activities have been criticised for not being sufficiently tailored to meet the needs of specific workplaces. However, these activities are now being modified to overcome this criticism through analysis of the extensive data they hold on health in the workplace to design more accurately targeted programmes. They profess to have a positive attitude towards workplace health promotion, and indeed it would seem to be in their interest to encourage such activities, as they carry a large part of the financial risk associated with health related absence from work.

Workplace actions for health have been an important issue for the German trade unions since the beginning of trade unionism in Germany, through their concern for the improvement of working conditions. Trade unions activities are mainly applied at two levels - improving the legal basis of OHS and using other instruments (only partially related to health). At the first level, the expanded bargaining power of trade unions and works councils is the most important instrument in influencing legislation. At the second level, such instruments as regulations on collective wage agreements, co-determination rules, and working time agreements are important.

Despite these instruments are being available, trade unions only rarely participate in discussions of occupational health promotion. They adopt the view that these discussions are primarily a marketing strategy of the sickness funds, while approaches like those of the WHO are seen as being too idealistic and have little to do with the reality of the workplace.

The improvement of the health of the workforce is not a very important issue for employers. Their over-riding concern is with sickness absence rates, which is seen as an important element in the competitive position of a company. However, proactive measures are only evident when there is a relative disadvantage in relation to a company's competitors, though there is a growing awareness among the employers of the need to improve working conditions for certain parts of the labour force such as women and older workers.

There are a growing number of companies with problematic sickness absence rates relating to either the whole company or sub-sections of it. Many of these companies are open to new approaches to the problem, and some go as far as to propose improvements in working conditions as a solution. More typically however, efforts are directed at improving existing OHS structures.

3.3.3 Attitudes and actions of key actors in Greece

The results from the Greek interviews (which were conducted with relevant government departments, employer organisations, trade unions and professional bodies) indicate that there is a wide range of workplace health activity taking place. All of these bodies expressed positive attitudes towards health promotion, but they expressed reservations regarding the extent of these activities.

Increased activity in the area is planned by many agencies, though much of this activity is confined to single health issues. Specific criticisms concern the failure of the State to adequately fund OHS infrastructure and developments, and a failure to implement existing legislation. A certain level of resistance to the concept of workplace health promotion was detected, though there was also evidence of a positive change in attitudes towards the concept. The need to increase general awareness both inside and outside of the workplace and worker participation was emphasised by many of the interviewees. Accordingly, much of the planned activity appears to consist of health education campaigns.

3.3.4 Attitudes and actions of key actors in Italy

The Italian study undertook a wide ranging series of intensive interviews which examined the views of both institutional actors and private and public companies. The Italian interviews used a slightly different protocol to the studies in the other countries. The first issue examined concerned the promotion of decisive action for health protection. Many respondents saw the value of adopting a multidisciplinary, multi-method approach to the problems of health and safety. However, resistance to the concepts was identified which took an active form of in the case of many companies, and a passive form in the apparent reluctance of many actors to instigate actions on their own behalf - many of their roles were dependent on others and were confined to information dissemination.

The second issue examined concerned interviewees awareness of examples of good practice in relation to workplace health. Most respondents were able to identify what they termed innovative actions in which they had been involved. However, the level of innovation in some of these actions may be open to question. Many of the initiatives appeared to be confined to single issues, and others focused on traditional H&S concerns.

A range of opinion was expressed regarding the interviewees attitudes towards workplace health promotion. The general feeling was that innovative actions are not sufficiently encouraged by the respondents organisations. The reasons for this predominantly negative orientation were varied, and had a pragmatic basis. However, some of the organisations have indicated their intention to play a more active role in the future.

In general, the Italian study revealed some good examples of workplace health promotion and identified generally positive attitudes towards the issue. However, problems of health culture and health priority in conjunction with a diffusion of responsibility between agencies has meant that the practice of workplace health promotion was hindered, with no agency having clear responsibility for the promotion of workplace health actions in the sense used in this research.

3.3.5 Attitudes and actions of key actors in Ireland

A total of eight interviews were conducted in the Irish study, with representatives of industry, government, trade unions, voluntary bodies and professional associations. These indicated that there is a very low base of activity taking place in Irish workplaces. At present, there is no single body responsible for workplace health promotion, and furthermore, there appears to be no great demand for it. However, some traditional OHS-type activities do take place, and there appears to be an increasing level and range of activities in existence.

A number of common themes emerged from the interviews. All of the respondents indicated the need for increased awareness of workplace health issues. Many respondents emphasised the importance of enforcement of legislation, which is currently seen to be less than perfect. In this regard, many respondents were unsure as to how the provisions of the new Irish health and safety legislation would work out in practice. A further concern to many respondents was that health actions may be used to the disadvantage of the individual worker. (The role of OHS professionals was often perceived to be compromised by the fact that they reported to management).

3.3.6 Attitudes and actions of key actors in the Netherlands

A total of four interview interviews were carried out in the Dutch study with representatives of two government departments, the employers organisations and the trade unions organisation.

Approximately one third of all Dutch employees are covered by an occupational medical service either provided by the company or the State. Currently there are 220 such services employing about 1000 doctors and 550 nursing staff. These services engage in a range of traditional activities in addition to some more innovative activities such as workstation research.

The attitudes of the major actors in the Netherlands can be illustrated by the deliberations of a tripartite working group set up to address the issues of work-related absenteeism and disability. Some of the innovative aspects of their deliberations include the recommendation of "integrated quality awareness" programmes, in which questions of technology, organisation and quality of work are considered. However further development of company medical services regarding these programmes has caused some disagreement, with employers preferring extension on a voluntary basis, while unions seek compulsory arrangements. The government hold the view that the social partners should reach agreement. Similar differences of opinion have occurred in relation to other issues. In general the attitude of the government is directed at strengthening the social partners own responsibility for safety, health and wellbeing in the workplace. The government decides the limits and is responsible for control, but at the same time financial incentives have been introduced to encourage employers (and employees) to improve quality of work.

3.3.7 Attitudes and actions of key actors in Portugal

Eleven interviews were held with key personnel in Portugal. These included representatives of the Ministry of Employment and Social Security, the Ministry of Health, a range of industrial associations (both employers and trade unions) and professional bodies.

In general, there was a low level of knowledge and awareness amongst these bodies about health promotion in the workplace. It is only in recent times that collective employment agreements have included clauses on health and safety at work and even then only in very general terms. Trade unions for example tended to view health and safety in the light of general collective bargaining procedures, and have typically sought payments or allowances for work involving hazards. In this regard, the unions feel that they are in a compromised position since if they seek to promote health and safety in its own right, their membership lose much needed income. In addition, the employers feel that it is easier to pay subsidies rather than to improve working conditions. In general there is no wide spread risk prevention culture in Portuguese companies.

Despite this unpromising background, there have been some initiatives in recent years, largely in the health education arena. These actions are not always being viewed in a positive light by workers, who sometimes fail to draw a distinction between the benefits of health education and the dangers of social control. The situation in Portugal is dramatically illustrated by the fact that the introduction of compulsory breaks during working hours is only beginning to appear as a clause in collective agreement regarding employment conditions.

There is no coherent health promotion policy in the workplace in Portugal. Some individual activities do take place which may have a positive affect on working conditions, but these have not been informed by a major philosophy or policy on the behalf of the actors concerned. However, it should be noted that in common with Greece, there are quite high levels of social activities provided by some Portuguese companies, at least by northern European standards.

In Portugal, both the employers and the trade union movement profess positive attitudes towards workplace health, but these attitudes appear to be rarely translated into practice. However, there does appear to be some change taking place in health and safety at work provisions. For example, inter-sector co-operation in occupational health is seen as desirable and necessary, but activities in this field are still uncommon. In relation to health education, the lack of resources, both financial and human have held back the development of activities in this arena.

Occupational medical services generally provide curative rather than preventive services. This state of affairs fits in with prevailing health culture, which has a curative rather than a preventive orientation. Furthermore, there is little pressure on occupational health services to change as, for example, many trade unions at company level do not tend to draw distinctions between prevention and clinical services.

3.3.8 Attitudes and actions of key actors in Spain

The aim of the interviews conducted in Spain was to describe occupational health education and promotion activities, who is responsible for them, their content, their target group, their underlying methodology, the results being achieved and their value. Face-to-face interviews were carried out with 41 respondents working for 40 separate agencies. These interviews provided a wealth of detail on the nature of occupational health activities carried out in Spain.

The interviews revealed a wide range of training activities, which varied in their methods and the kinds of trainees they were targeted at. Courses for trade unionists, safety delegates, occupational nurses and physicians, and engineers were described. Trade unions and others appear to have adopted Italian models of training on a fairly wide basis, particularly in relation to risk assessment methods. Some problems with attendance and the subsequent application of course material have been noted. The model of Occupational Health Assemblies (occupational health centres providing information to safety delegates from companies for discussion) employed by some occupational health centres seems to be particularly successful.

Information campaigns tend to embrace a wider public rather than the restricted target group of trade-union delegates aimed at by most training programmes. The majority of institutions consulted produce magazines, posters, special reports and other information material on a regular basis. Information campaigns are nonetheless a way of reaching and raising the awareness of population groups exposed to specific risks.

Risk analysis measures are not as prevalent as training courses. They seem to occur on a relatively haphazard basis, and have met with limited success to date.

In summary, the common features of workplace health promotion in Spain include:

- a low level of activity;

- most actions consist of health-education and training

- teaching methods tend to have a technical focus, with a marked medical content and little student participation;

- health promotion actions tend to experimental;

- in general, actions tend to be of a medical nature, using the techniques of medical consultation and relying on the company doctor;

- medical actions are primarily aimed at monitoring risk factors associated with cardiovascular diseases, malignant tumours, the effects of alcohol abuse and stress: smoking, alcohol consumption, diet, obesity, physical exercise, stress, etc. In general, actions are designed to monitor one or a small group of these factors;

- there is confusion between regular medical examinations, screening and primary prevention;

- very few employers encourage this type of action. Workplace actions tend to be undertaken at the insistence of specialists concerned to ensure that workers are provided with a useful and effective company medical service and by employers' mutual-insurance associations.

3.3.9 Attitudes and actions of key actors in the United Kingdom

Interviews were carried out with six key actors in the UK occupational health scene. These included employers, trade unions, the government occupational health agency, local government occupational health agencies, and a representative commercial firm of caterers who would have the potential to influence the health of the workforces of their customers.

Each of the organisations were asked to identify their most important health priorities. Both the TUC (Trade Union Congress) and the CBI (Confederation of British Industry) emphasised the linkage between health and safety issues. The HSE (Health and Safety Executive), Environmental Health Departments and the Institute of Occupational Health all emphasised the importance of addressing safety issues as a priority particularly in the area of preventing accidents and industrial diseases.

Specific health issues of concern included: stress and mental health (the CBI, TUC); alcohol and drug misuse in the workplace (TUC); alcohol, smoking and HIV infection (several organisations).

All of the organisations interviewed emphasised the primacy of the need for providing safe working conditions and basic occupational health care. The majority also recognised that the workplace is an appropriate setting for addressing wider health issues. In general, these organisations are willing participants, but felt the need for guidance from health promotion organisations.

In some organisations, action for health had been introduced to meet other objectives, for example, recruiting members to trades unions, or as a new marketing promotion for the catering company. This is a valuable approach, as it serves to produce benefits not only in health terms but also in other areas important to the organisation.

Other common themes in the interviews concerned the need for reorientating and extending of existing occupational health services; the need for significant changes in the training of occupational health staff; and the need for evaluation if benefits are to be made clear to all of the parties concerned. Finally, it was clear that differing views were held on the relative merits of statutory versus voluntary action in the area.

3.4 So what is the Official Story ?

There is difficulty in summarising such a wide range of interviews from such a diverse number of contexts. However, a number of common themes emerged

- In all eight countries there is a low level of actions which could be classed as genuinely innovative. Furthermore, many of the studies pointed to low levels of activity in more traditional H&S activities. Some studies also pointed to the fact that relatively high proportions of actions were not successful (e.g. Germany, Spain) with regard to really improving the health of the workforce.

- Many studies pointed to the difficulty of identifying a single agency with responsibility for workplace health actions. This fragmentation of responsibility appears to have be reflected in fragmented service provision. While this state of affairs is especially true in the case of workplace health promotion, it is also largely true of traditional H&S actions.

- An implication of many of the interviews was that health promotion would not take place unless it was either compulsory and enforced, or economically beneficial to do so.

- Many organisations emphasised the strong links between safety and health issues on the one hand and between organisational indicators of efficiency, e.g. absenteeism, morale on the other.

- Many organisations (both trade unions and employers) stressed the importance of participation by workers in health initiatives if they are to be successful. Related to this point is the need to ensure that health actions are well targeted if they are to have high levels of uptake.

- Awareness of both general and workplace health issues was judged to be almost uniformly low in the eight countries. A large number of respondents emphasised the need for major awareness raising programmes.

- The utility of programmes which aimed to modify individual behaviour was questioned. Trade unions in particular were wary of their use.

- Trade unions reported a certain degree of scepticism in relation to workplace health actions. They sought to ensure that the conditions of their implementation would not compromise their members.

- Many of the programmes in existence confined themselves to single health issues.

- There is currently in many countries a debate on the relative merits of statutory versus voluntary action in relation to workplace health action. The Dutch and UK legislation in particular seeks to promote voluntary action. A related point concerns a move away from detailed regulations and standards towards a process whereby the process of workplace health protection is legislated for.

- Respondents in many countries pointed to the inadequacies of the implementation of legislation. The cause of workplace health actions would be considerably advanced if existing provisions were enforced.

- There was also a feeling that the legislative contexts in some countries were inadequate

- The resources available for health promotion were seen to be inadequate both in terms of infrastructure and finance in many countries. This lack of resources may have placed undue pressure on workplace health initiatives to be self-financing.

- Governments are generally in favour of workplace health actions, be they innovative or otherwise. However, the lack of resources indicates the weight of their commitment to this ideal.

- Health agencies are also generally committed to the ideal of innovative workplace health promotion. However, they do not always see themselves as being responsible, nor do they necessarily have the resources to prioritise this activity.

- Trade unions are also in favour of innovative workplace health actions. However, a lack of awareness amongst their membership means that in practice there is little commitment to the concept. The training and awareness programmes in which many are involved may rectify this situation ultimately.

- Private sector organisations tend to be market led in their activities.

4. The Reality of Workplace Health

This chapter contrasts the positive attitudes expressed by the major actors with the practice of workplace health promotion as described in the responses to the survey. In particular, it describes the samples from which data was gathered, the health characteristics of the companies involved, the levels of health activity the levels of participation by the major actors in health actions and the process whereby health actions are established. In brief, levels of health activity are quite low, are largely confined to traditional activities, and are in marked contrast to the positive attitudes professed by key actors in the course of the interviews described in the previous Chapter.

Data on the extent of workplace health activities is not readily available in most European countries (Wynne, 1990), and there is even less data available on workplace health promotion. The current programme of research provides a rare opportunity to acquire such data in many countries, and more importantly, provides a unique opportunity to generate survey data which is comparable from country to country through the use of a common questionnaire.

4.1 Questionnaire development

In drawing up the questionnaire (see Appendix) four different sections were developed. The first sought company demographic details, the second sought information about the nature of health actions which were taking place. The third looked at the mechanisms for establishing workplace health actions, while the fourth looked at companies plans and priorities for the future with regard to health in the workplace.

Development of the questionnaire was a cooperative activity undertaken by all seven of the national researchers (the survey was not undertaken in Portugal). Initially a draft questionnaire was drawn up, which addressed the areas of company demography, health activities, the establishment of health activities and plans for the future. This questionnaire was then piloted in each of the countries and appropriate modifications were made on the basis of the results from these pilot studies.

The end product was a core questionnaire, which contained identical questions in each of the countries. In some countries extra questions were inserted intothe questionnaire on the basis of the research interests of the individual research group.

By and large, the questionnaire was equally applicable across all of the 7 countries. However, there were a number of items on the questionnaire, which tended to address redundant issues in the different countries. For example, in Germany health and safety committees are compulsory, and in Spain, the use of the word 'staff' in the section on participation was taken to mean occupational health staff. However, the differences interpretation which commonly occur in cross national studies as a result of cultural and legislative differences were minimised in the development of the survey instrument.

4.2 Sampling

In all 1451 organisations responded to the survey which took place in seven of the EC countries (Germany, Greece, Ireland, Italy, the Netherlands, Spain and the UK). It should be borne in mind that in the drawing up the various national samples, no attempt was made to achieve representativeness. (In general, it is extremely difficult to draw up accurate samples of workplaces in any or all EC member states). In practise many different sampling strategies were used, Business directories and in-house databases were the predominant sources from which the samples which were drawn.

In practise the only stipulations in drawing the samples concerned the aim that they should be drawn from at least two regions within the countries concerned and that at least 200 responses should be obtained to the questionnaire. In most cases, these were selected to come from areas of heavy concentration of industry and more rural areas.

4.3 Response rates

The response rate to the survey varied between the countries, from approximately 11% to approximately 35% (see Table 4.1), with an overall average of 23.6%. This level of response is quite low and is not untypical of surveys of industry (at least according to the researchers own experience of conducting similar work in their own countries).

Table 4.1 Response rates in the different countries

	Q'aires issued	No. of returns	Response rate(%)
Germany	1419	161	11.3
Greece	535	200	37.4
Ireland	804	138	17.2
Italy	725	202	27.9
Netherlands	803	207	25.8
Spain	1215	312	25.7
UK	656	231	35.2
Total	6157	1451	23.6

These low response rates are likely to be a reflection that companies and organisations which are relatively active in the field of health in the workplace were more likely to respond to the survey than those which were inactive in this area. It is therefore it almost certain that the results which are presented in this and the next chapter are not at all typical of levels of activity which take place in the wider population of companies in each of the countries. However, this is not a drawback, given the purpose of the study in the first instance. The major interest in this study was to assess the kinds of health activity which are taking place, rather than their absolute prevalence and of more interest was to examine how they are established and how they are organised rather than seeking to make definitive statements about population characteristics of workplace health action.

4.4 Demographics of the sample

Across the seven countries 18.4% of the sample operated in the public sector, 75.5% were private sector companies, while 6.1% classified themselves as 'other' (these largely refer to semi-state companies and co-operatives). 27.2% of the organisations in the sample were multinationals with almost 73% being owned within the boundaries of the member states concerned.

N.A.C.E. Divisions

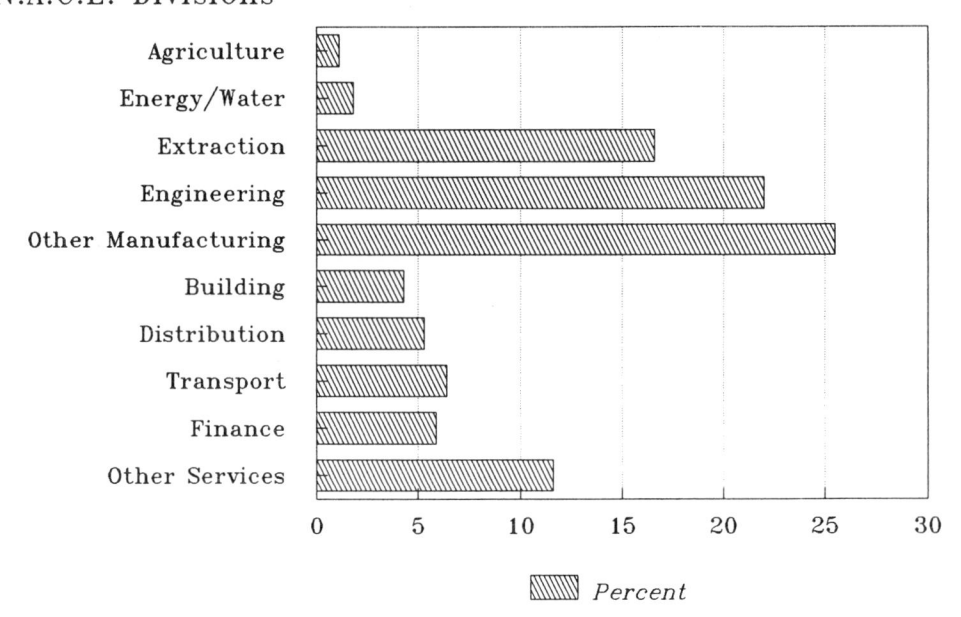

Figure 4.1. Industrial sector of respondent organisations

The area of business in which the sample operated indicated that the majority were involved in manufacturing (the NACE codes were used for classifying types of businesses). In all 25.4% came from 'other' manufacturing industries. 22% from metal manufacture, mechanical, electrical, and instrument engineering. 16.5% from extraction and processing industries. 11.4% from other services. The remainder of the sample came largely from transport and communications, banking and finance distributive trades and building and civil engineering sectors (See Figure 4.1). The biggest single sectors concerned, chemical industry, metal manufacturing and the textile industry where more than 5% of the sample was represented from these sectors.

This breakdown of respondent organisations by sector indicates that the sample was biased towards the manufacturing sector, while the services and agricultural sectors were under-represented. This finding supports the view that the sample contained a preponderance of companies who are active in the health and safety field, as the manufacturing sector has perhaps the longest tradition of activity in these areas.

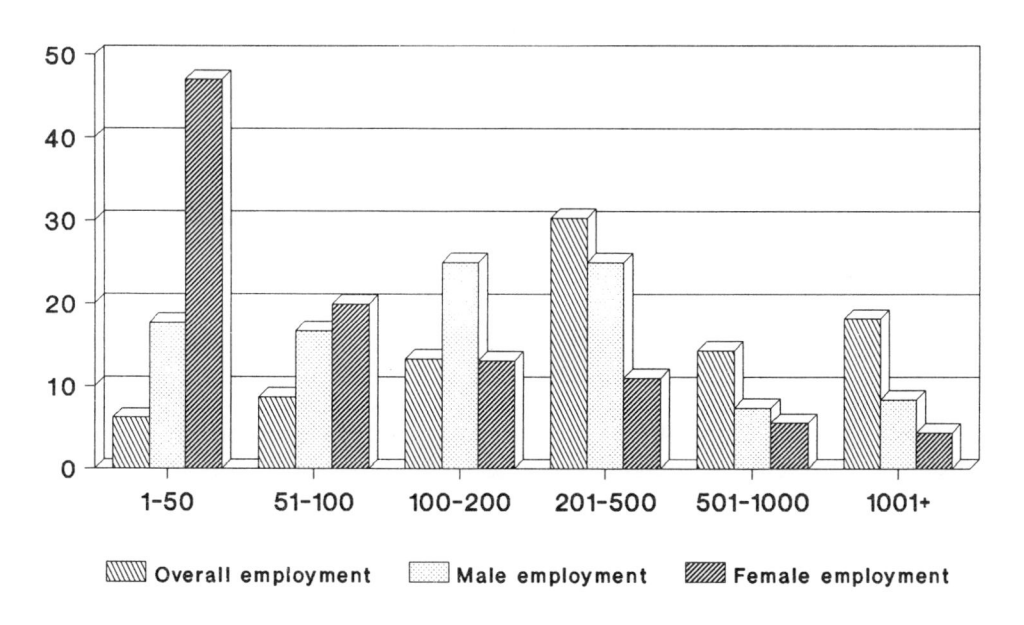

Figure 4.2. Numbers employed in respondent organisations

Figure 4.2 depicts the size of the respondent organisations in terms of their number of employees. From these figures (which relate to total employment, and total male and female employment) it is clear that the size of the respondent organisations tended to be very large in comparison to the average size of enterprises in most countries.

There was quite a high level of variation in levels of trade union membership within the organisations. As few as 6.2% organisations had no trade union membership, while a third (33.3%) had a trade union membership between nought and twenty five percent of the workforce. 17.6% of organisations had between 25% and 50% of their workers as members of trade unions, while 19.1% had between 50% and 75% trade union membership. Finally, 30% had between 75% and 100% trade union membership among the workforce. These figures indicate that the sample was biased towards the older industries where trade union membership levels tend to be higher.

Also of interest were the kinds of premises which respondent organisations operated. 65.2% operated office premises, 75.5% operated production or manufacturing areas, 44.7% operated public areas, while 31% operated 'other' areas. These findings again indicate that the sample was biased towards the manufacturing sector.

4.5 Company Health Characteristics

Information was obtained from the questionnaire regarding the infrastructure available within companies to engage in health and safety activities. The infrastructure in question concerned the presence of safety oriented entities, such as occupational health departments and health and safety committees on the one hand, but also with more health oriented indicators, such as the presence of health policies and health budgets. It should be noted that these later indicators specifically referred to health rather than safety policies and budgets.

60.6% of companies had occupational health services on site which had qualified or medical or nursing staff working in them. 70.2% had a health and safety committee or equivalent (it should be noted that in some countries, e.g. Spain, these are compulsory for enterprises of over a given size). The membership of these committees yielded interesting insights, with 68% having staff representatives on the health and safety committee, 68.3% having management representatives on the committees, 55.5% having trade union representation, 62.1% having occupational health staff representation, and 34% having other members. (The membership of these committees is sometimes laid down by law).

Respondents were also asked to nominate what they considered to be the most important health problems among their employees. (Note: they were not asked to distinguish between health problems of occupational or non-occupational origin). Aggregation of the data across national boundaries was not possible in this instance, as the differences in coding of the responses to this question were too great. However, Table 4.2 reproduces the data from each of the seven national reports to enable comparisons to be made.

Table 4.2 Health Problems of employees

	Netherlands	UK	Germany	Italy	Spain	Greece	Ireland
Locomotor		50%	NA	55%		32%		17%
Stress		45%		22%		14%	13%	31%
Cardiovascular disease		<12%		28%		5%	20%	12%
Infections		<12%						
Private life		<12%						
Respiratory				28%		26%	10%	
Alcohol				16%		7%		
Noise/hearing loss				11%	21%	10%		
Toxic substances					10%			
Occupational injury/disease					10%	10%		17%
Orthopaedic/rheumatic							25%	
Hearing/vision							12%	
Lifestyle								59%
Environmental problems						4		26%

Comparison of the most prevalent health problems from various countries is difficult given the different interpretations placed on what exactly constitutes an illness and also the different groupings applied by the national researchers. However, it is fairly clear that problems in the locomotor system, be they musculoskeletal or orthopaedic rate highly in most countries. The issue of stress also appears as an issue in many countries, with the exception of Italy, with the highest proportions occurring in the Netherlands and in Ireland.

In general, whatever the classification systems applied by the national researchers, it is clear that many of the health conditions or health problems sited are very much amenable to work site health promotion, even though many of them may not have direct links to hazards present in the workplace.

The results from the current study can be compared with the recently published Eurobarometer study, (EC 1992), which asked a thousand representative respondents from each of the EC member states questions about health risks in the workplace (see Table 4.3).

It should be noted that there are significant differences in the methodologies used in the current study and the Eurobarometer survey. In the current study health problems were investigated using an open ended question format, while in the Eurobarometer survey a closed or fixed response format was used. Nonetheless the data produced by Eurobarometer provide some interesting comparisons. Perhaps the most striking difference concerns the issue of stress where, in many countries it was one of the most prevalent health risks to be found in the workplace, more so than problems relating to breathing, muscular pains, eye problems or even backache. It is also of interest to note that stress as a problem appears to be much more significant from the findings of the Eurobarometer survey than it does from the current work. This may be partly due to the often noted tendency of fixed response questions to produce higher rates of prevalence than open ended ones. In any event, the level of problems cited in the Eurobarometer survey also support the thesis that the work site is a particularly appropriate arena for health promotion activity.

Table 4.3. Health risks in the workplace - Findings from Eurobarometer survey

	Netherlands	UK	Germany W	Germany E	Italy	Spain	Greece	Ireland
Breathing	22		21	20	26		33	24
Muscular pains	19	24	31	35	32	39	54	
Eye Problems		34	27	26		26		25
Tiredness	22	24			28	27	67	27
Backache	44	38	58	53	39	42	44	26
Stress	41	56	52	45	45	28	52	46
Others	57	68	56	45	62	54	71	52

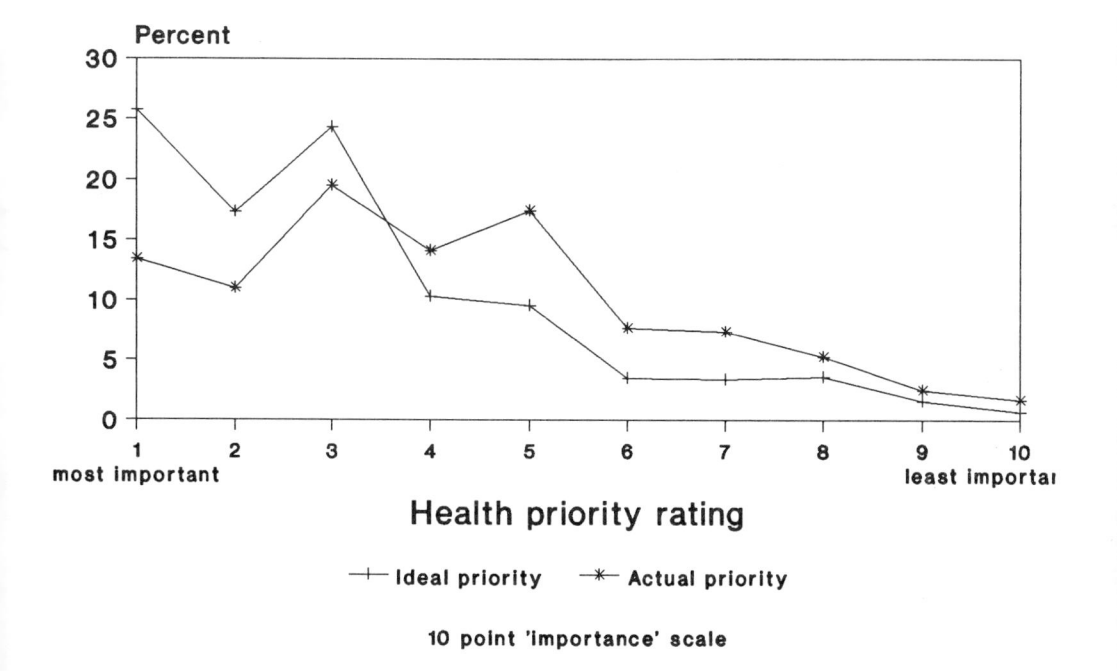

Figure 4.3 Relative health priorities of the respondents

An attempt was made to assess the level of priority which respondents in organisations gave to health as an issue in company operations. They were asked two questions, one which asked what would be the ideal priority for health, rated on a scale of one to ten (where 1 = highest priority and 10 = lowest priority), and secondly, what is the actual priority given by the company to health. It should be noted that these are crude measures that to some degree depend on the individual respondent to the questionnaire, who may or may not have sufficient information to give an accurate view of what these priorities might be. Nonetheless, as might be expected health was generally rated as being quite important, with the mean score on the ten point scale being 3.2. However, when it came to the actual priorities given to health, health seemed to be much less important with a mean score of 4.2. These findings are illustrated in Figure 4.3.

Respondents were also asked if the company had an explicit or written health policy, as opposed to a safety policy. In all, 31.5% of respondents claimed that their companies had written health policies. People who were involved in the formation of this policy, included managers in 96% of cases, occupational health staff in 79% of cases, staff representatives in 77% of cases, trade union representatives in only 39% of cases, health and safety representatives in 74% of cases and external consultants in 48% of cases. Furthermore, 63% of companies stated that they were likely to further develop their health polices or to develop one if they did not already have one.

These findings are significant as the presence of a health policy indicates that enterprises have given some thought and effort to the issue of health. While the presence of health policies may not be indicative of actual activity in the area, they at least signify that health is on the agenda of companies.

A harder measurement of an enterprise's commitment to health concerns the presence of a specific health budget. As many as 32% of companies claimed to have such a defined health budget. Again this refers not to safety budgets, but to a budget specifically designated for health purposes. 55% of respondents believed that the amount in the budget would increase in the future, whereas only 3% believed that it would decrease and 42% that it would stay the same.

These results indicate that it is almost certain that the sample is atypical in terms of the level of health infrastructure in company health characteristics as compared to the general population of companies in the seven countries. The numbers of companies with budgets and the numbers of them with explicit health policies strongly suggest that the current sample is largely made up of companies who are already active in the area of workplace health.

4.6 Health actions in the workplace

Figures 4.4 to 4.8 illustrate the percentage of workplaces in which basic health actions and health actions in which improving the health of the workforce was a consideration took place. (These 30 actions operationalise the concept of workplace health promotion in the survey). The most notable feature of these figures is that many actions which can have an influence on health are not perceived as having health improvement as a major stimulus in their taking place. This occurs even with activities which would appear on the surface to be almost solely concerned with health, for example, health screening for all staff, health screening for at risk staff or introducing exercise facilities. This finding has implications for the marketing and establishment of workplace health activity - health actions obviously take place for reasons other than health improvement, and if these other reasons are satisfied then the likelihood of their establishment must increase.

The most prevalent class of activities, concerned actions directed at safety and the physical work environment where in excess of 60% of respondent organisations claimed to have introduced machine guards, protective equipment, improvements in lighting, heating, ventilation and noise reduction. (It should be noted that many of these actions are the subject of legislation). The least prevalent of actions prior to some of the organisational interventions and the social welfare services actions, in particular, actions such as shift design, stress control programmes, support programmes, job design tended to take place with low levels of frequency. By contrast to safety actions, actions in the welfare and organisational arenas are least subject to legislation.

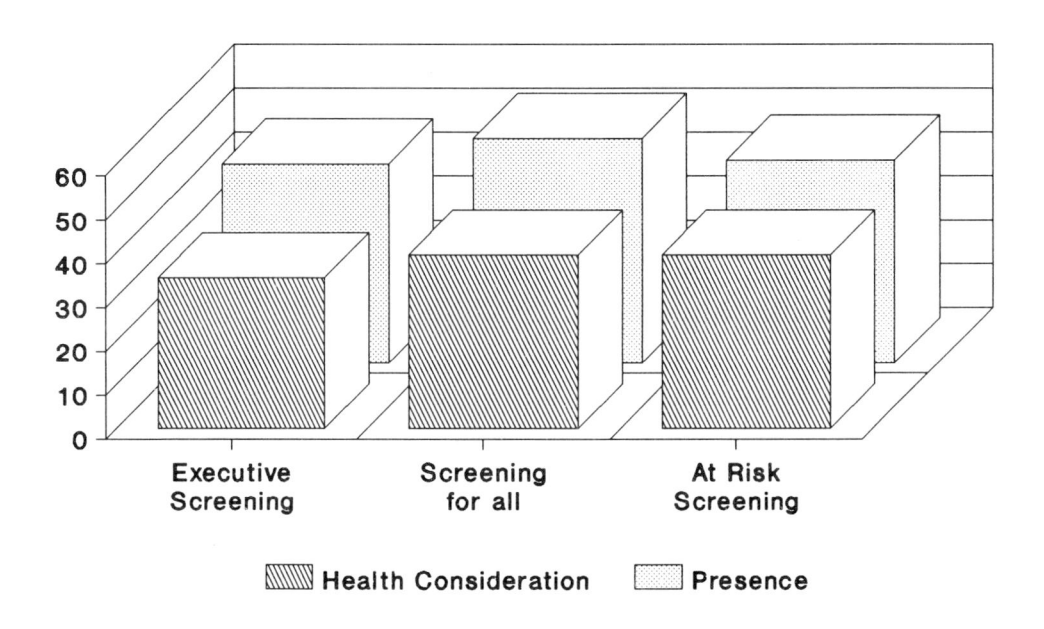

Figure 4.4 Prevalence of health screening activities

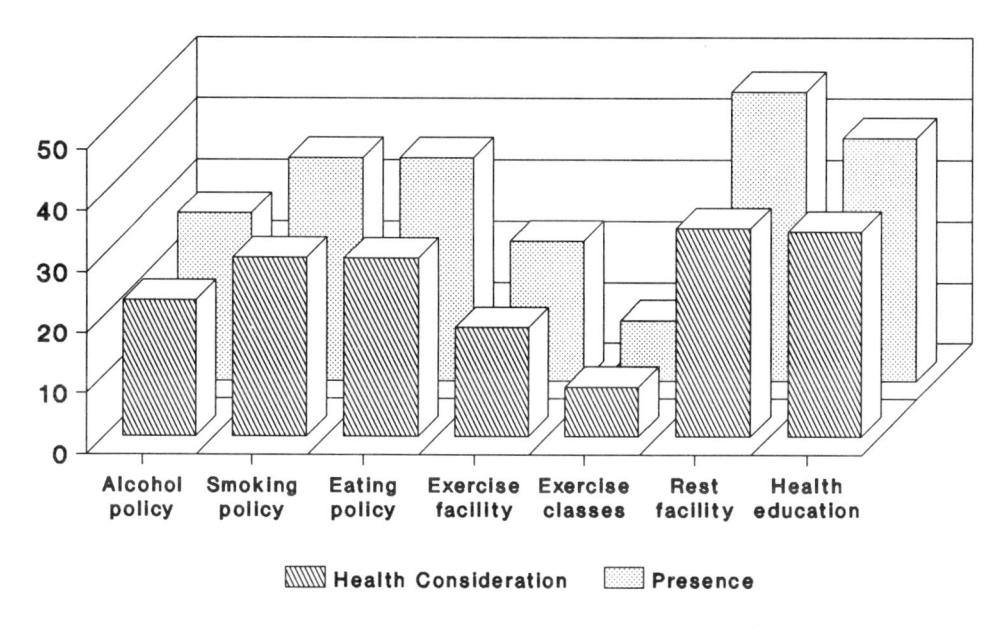

Figure 4.5 Prevalence of healthy behaviour activities

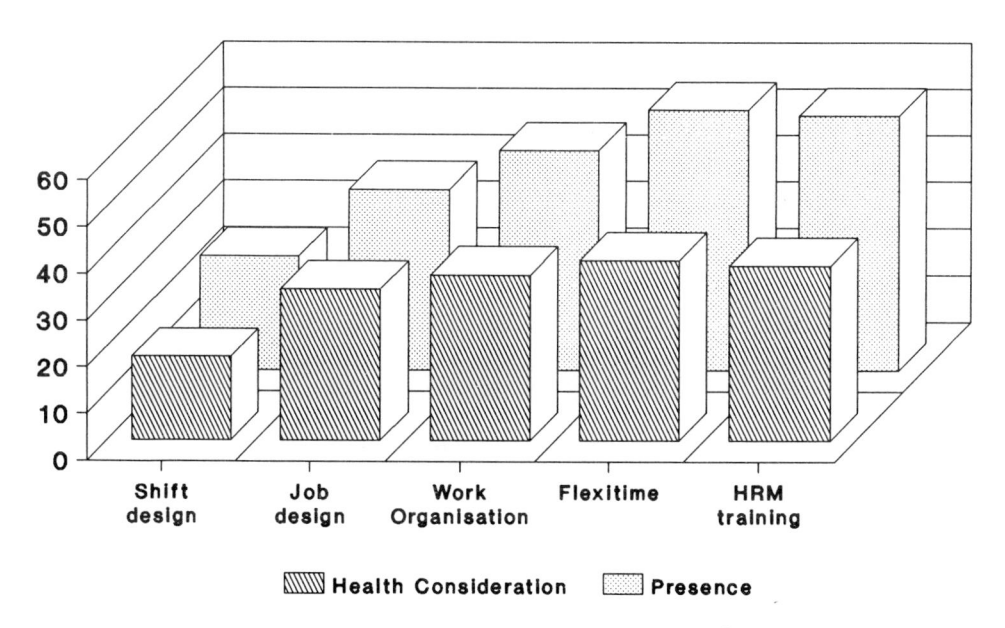

Figure 4.6 Prevalence of organisational interventions

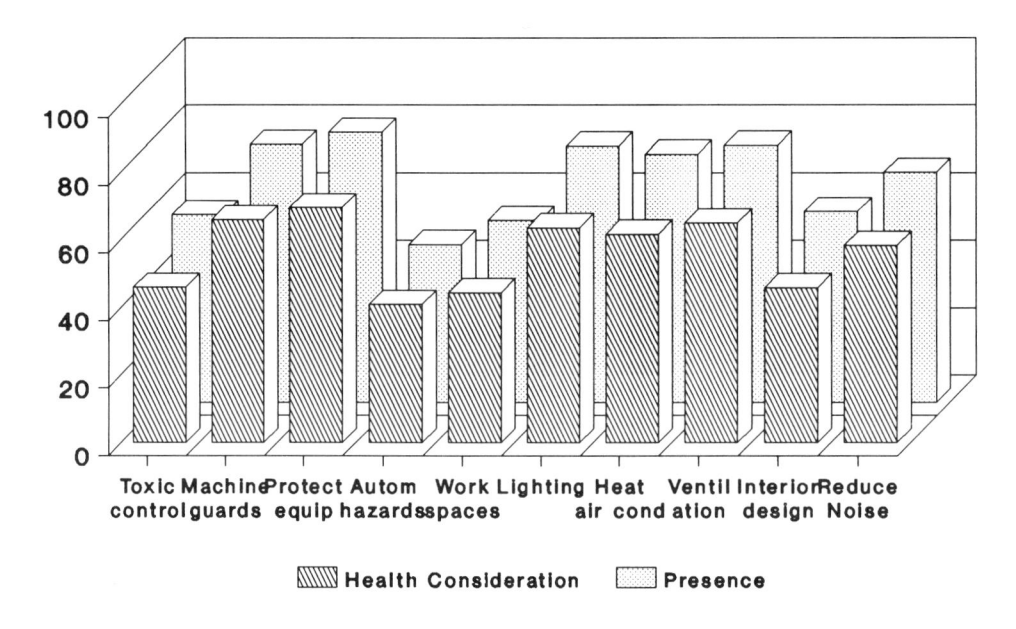

Figure 4.7 Prevalence of safety/physical environment activities

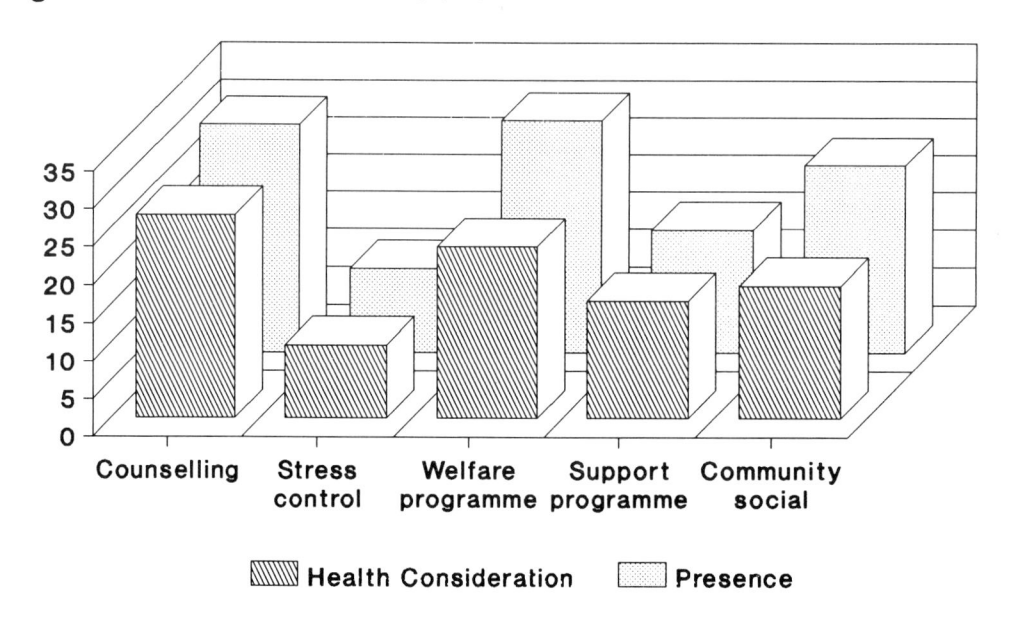

Figure 4.8 Prevalence of social/welfare activities

Table 4.4 examines the issue of the proportion of activities which have health as a major consideration in their taking place. Some interesting findings emerge when these are examined, as the higher the proportion which have health improvement as a prompting factor, the 'purer' that activity is in the health sense. Furthermore, where low proportions of activities have a health component, it indicates that these actions are taking place for reasons other than health improvement. It may also indicate a gap in awareness of the health improvement potential of that activity. The policy implications of these findings are quite clear - in cases where gaps in awareness are high information/education interventions are indicated.

In the Table the prevalence of activities is given, as is the prevalence of activities which have a health component. The ratio (a 'health index') between these two indices is then calculated - a value of 100 indicates that activities took place solely for health improvement reasons and that there is no gap in awareness of the health improvement potential of that action, while a score of 0 indicates that activities took place for reasons other than health improvement. In this situation a sizeable awareness gap may be present.

Table 4.4 indicates that the activities which had the lowest health improvement component in their taking place were the flexitime programmes, community or social programmes, Human Resource Management Training, the provision of rest, social or shower facilities, welfare support programmes, the design of shift schedules, work organisation programmes, executive screening programmes and the provision of exercise facilities. In the case of each of these activities between 20% and 30% of them take place for reasons other than health improvement. It is obvious from this list that these other reasons are largely concerned with welfare programmes and with operational concerns of work design and scheduling. However it is interesting that these kinds of activities are not seen as been relevant to health improvement - perhaps an awareness gap genuinely exists in relation to these activities.

Table 4.4 Proportion of activities which have health as a consideration

Health Activity	% taking place	% with health consideration	Health Index
Executive screening	45.3	34.5	76
Screening for all	51.1	39.7	78
At risk screening	46.2	39.7	86
Alcohol policy	27.6	22.6	82
Smoking policy	36.6	29.5	81
Healthy eating policy	36.6	29.4	80
Counselling support	30.2	26.8	89
Exercise facilities	23.0	18.1	79
Exercise/lifestyle classes	10.1	8.2	81
Rest/social/shower facilities	47.3	34.3	73
Health education	39.8	33.8	85
Shift schedule design	24.3	17.9	74
Stress control	11.0	9.7	88
Job design	38.7	32.3	83
Work organisation	47.1	35.4	75
Working time flexibility	55.9	38.5	67
Welfare support	30.8	22.6	73
Support programmes	16.1	15.5	89
HRM training	54.6	37.4	68
Community/social programmes	24.8	17.5	71
Toxic substance control	55.6	46.0	83
Machinery guards	76.4	65.9	86
Protective clothing/equipment	80.0	69.5	87
Automating hazardous processes	46.8	41.0	88
Individual workspaces	53.9	44.4	82
Lighting	75.9	63.6	84
Heating/air conditioning	73.5	61.7	84
Ventilation	76.1	65.0	85
Interior design	56.5	45.9	81
Noise reduction	68.1	58.4	86

It is perhaps surprising that executive health screening is also contained in the list of activities with the lowest health index. The other reasons for executive health screening take place may relate to its reputation as being a fringe benefit for executives.

The activities which had the highest health index were largely concerned with safety or physical work environment interventions such as the provision of machine guards, toxic substance control and the automation of hazardous processes. In addition the activities of health screening of at risk groups, stress control and support programmes (e.g. Alcoholics Anonymous, Gamblers Anonymous) also rated highly on the health index. The reasons for these high ratings are largely related to an equating of the issue of safety and health, and to the provision of semi-clinical services.

4.7 So what is the reality of workplace health ?

It is clear that notwithstanding the fact that the sample is largely made up of companies who are already active in the health arena, that the most prevalent activities in these workplaces are largely concerned with safety and interventions to the physical work environment. The least common activities are concerned with organisational interventions.

It is also clear that many of the activities examined in the questionnaire take place for reasons other than health improvement. These other reasons are largely concerned with fulfilling the provisions of legislation and to the provision of fringe benefits. In addition many organisations see the need to become involved in support programmes which have a semi-clinical orientation.

These finding indicate that there may well be a gap in the awareness within enterprises of the potential of many activities to influence health. This is particularly seen in relation to some of the organisational intervention activities such as shift schedule design, work organisation and job design, where even the relatively few activities which take place rarely occur with health improvement as a consideration.

5. What Actually Influences Workplace Health Actions?

This chapter describes the factors which are most significantly associated with the establishment of workplace health actions. In particular it explores the relative importance of participation levels, company demographics, health characteristics and the factors which prompt health activity with a view to identifying the key factors in establishing health actions.

5.1 Introduction

The analyses of cross-country data is always difficult, given the differences in language, culture and levels of economic development. It is made more difficult in the present case because of some differences in the samples of organisations drawn in the seven countries. However, there is a strong case for analysing the data at a European level for a number of reasons. Firstly, the data were not collected for purposes of describing a representative position, rather to gain insights into how workplace health actions become established. Secondly, even though there were some differences between the samples drawn in the different countries, these differences did not warrant the separation of the data into their national samples.

A number of issues are examined in this chapter. Specifically:

- to what extent are there national differences in the prevalence of
    - health activity
    - health activity in which health improvement was considered

- to what extent do company demographic factors influence:
    - health activity
    - health activity in which health improvement was considered

- to what extent do company health characteristics influence:
    . levels of health activity
    . health activity in which health improvement was considered

- which prompting factors most influence:
    . levels of health activity
    . health activity in which health improvement was considered

The data analysis was carried out in two stages. The first consisted of a bivariate analysis which examined the associations between the dependent variables and the independent variables. Of specific interest here were the effects associated with country, company demographics, health characteristics and the prompting factors on the twelve health activity variables and the six participation variables. (A complete listing of the dependent and independent variables is to be found in Table 5.1).

Table 5.1. Dependent and independent variables used in the analyses

Independent variables	Dependent variables
Company demography	*Health activities*
Size	Health screening 1
Sector	Health screening 2
Ownership	Promoting healthy behaviour 1
Trade Union membership	Promoting healthy behaviour 2
Premises	Organisational interventions 1
	Organisational interventions 2
Health characteristics	Safety/Physical work environment 1
Occupational health department	Safety/Physical work environment 2
Health and safety committee	Social/Welfare services 1
Health policy	Social/Welfare services 2
Health budget	Total health activity 1
Health priority	Total health activity 2

Table 5.1. Continued. Dependent and independent variables used in the analyses

Independent variables	Dependent variables
Prompting factors Legislation Personnel Problems Health Problems Staff Morale Absenteeism Productivity/Performance Staff Turnover Industrial Relations Public Image Accident rates Country	*Participation variables* Management Staff Trade union Health and safety representatives Occupational health staff External consultants

Note 1: The number 1 refers to health activities which have taken place, and 2 refers to activities which have taken place which have had a health component in their establishment.

Note 2: The factors which predict the participation variables are explored in Chapter 6.

The second stage of the analysis consisted of a series of multivariate regression analyses. This statistical technique allows for the simultaneous examination of the effects of the independent variables. It provides answers to questions concerning the relative importance of independent (or predictor) variables, e.g. is company size or the presence of an occupational health department more important in predicting levels of health activity?

The purpose of these analyses was to control for all significantly associated variables from the first stage - to seek the unique associations between independent and dependent variables. An additional interest in these analyses was the extent to which levels of participation in health activities by the six major workplace actors was associated with levels of health activity.

For reasons of space the details of the first stage of the analysis are not reported, with the exception of those carried out on the country variable, as this variable was of interest when considering whether to integrate the national databases or not. The other findings from the first stage of the analysis indicated that most of the demographic and health characteristics variables were associated with many of the dependent variables. This is probably due in part to the large sample size, even though a stringent criterion of statistical significance levels (0.01) was used.

5.2 Were there national differences in the samples ?

The issue of national differences in the data from the seven member states is of importance, not so much because of interest in examining different parameters of health actions in the workplace in the different countries, but in relation to the further analysis of data which is presented in the next chapter. It is important that the issue of country does not confound results from the various individual countries.

Not surprisingly there are some large differences between the countries in relation to the data discussed above. For example, in relation to the sector in which the respondent companies operated, Holland, Ireland the UK and Greece all had in excess of 20% coming from the public sector, while Germany had less than 10% from the public sector. In addition Holland had 15% and Greece 11% coming from the 'other' sector (which is largely made up of semi-state and co-operative organisations), whereas both Spain and Germany had less than 1% from this sector (see Table 5.2).

Table 5.2 Economic sector of the sample by country

	Public	Private	Other
Ireland	24.3	71.3	4.4
Spain	10.3	89.1	0.7
UK	26.7	70.1	3.2
Germany	9.5	89.9	0.6
Italy	12.6	79.0	8.4
Holland	27.1	57.6	15.3
Greece	20.5	68.5	11.0
Europe	18.4	75.5	6.1

There are also some marked differences in whether or not companies are multinationally owned (see Table 5.3), with as few as 12% and 13% multinationals in the Dutch and Italian samples while more than half of the German sample companies were multinationals (it should be noted that German companies in the sample were very large and that many of these multinationals may in fact be German multinationals).

Table 5.3 Ownership of enterprises in the sample by country

	Multinational	National
Ireland	29.4	70.6
Spain	33.3	66.7
UK	30.9	69.1
Germany	52.3	47.7
Italy	12.4	87.6
Holland	13.2	86.8
Greece	23.0	77.0
Europe	27.2	72.8

The most marked differences were seen in relation to company size (in terms of number of employees), with for example almost 40% of Irish companies in the lowest categories of size and none of the German companies in this category (see Table 5.4). At the other end of the scale 61% of German companies are in the largest size category while less than 10% were in this category in both Spain and Italy. These differences in size reflect the real situation obtaining in these countries, where for example the size of enterprises in Ireland is much lower than the EC average.

Table 5.4 Company size in the sample by country

	<100	100-200	201-300	301-500	501-1000	>1000
Ireland	39.5	20.9	8.5	12.4	7.8	10.9
Spain	24.3	25.4	16.2	13.4	12.7	8.1
UK	12.6	20.6	18.4	11.7	13.9	22.9
Germany	0	6.3	5.7	7.6	18.9	61.6
Italy	21.0	16.0	14.0	19.5	21.0	8.5
Holland	3.4	36.3	15.2	18.1	11.3	15.7
Greece	6.0	25.0	24.5	20.0	14.0	10.0
Europe	14.9	22.2	15.4	14.9	14.3	18.2

Number of employees

Levels of trade union membership also varied between countries, with very low levels of trade union membership reported from the Spanish and Dutch samples in particular (see Table 5.5). Very high levels of trade union membership were seen in Greece, Ireland, the UK and Germany, though this is probably not typical of levels of trade union membership amongst all companies in these countries.

Not surprisingly there were also significant differences between the health characteristics of companies in the different countries (see Table 5.6). As many as 85.6% Spanish companies, 81% of Dutch companies and 67% of German and Greek companies had occupational health departments, whereas only 22.5% of Irish companies had such departments. The high prevalence of occupational health departments in Spain and Greece relates to legislative provisions, while the relatively low figures in the UK and Ireland reflects the voluntaristic provisions for occupational health departments in these countries.

Table 5.5 Levels of trade union membership in the sample by country

	0-25%	26-50%	51-75%	76-100%
Ireland	28.3	6.7	15.8	49.2
Spain	63.4	19.0	13.0	4.6
UK	28.5	15.5	21.8	34.2
Germany	17.2	16.1	27.6	39.1
Italy	7.9	28.4	36.3	27.4
Holland	60.6	21.8	11.8	5.9
Greece	13.7	10.9	10.3	65.1
Europe	33.3	17.6	19.1	30.0

Table 5.6 Company health characteristics of the sample by country

	OHD	HSCOMM	Health priority High	Med	Low	Health Policy	Health Budget
Ireland	22.5	66.0	38.4	32.0	29.6	23.1	23.9
Spain	85.6	91.2	50.7	31.6	17.7	32.0	33.6
UK	34.5	76.6	29.1	34.0	36.8	25.8	26.9
Germany	67.3	88.2	24.7	45.3	30.0	17.0	27.6
Italy	48.0	41.5	60.4	24.4	15.2	29.2	35.6
Holland	81.0	56.2	30.5	39.5	30.0	39.4	27.2
Greece	67.4	65.0	67.0	15.6	17.3	49.0	47.8
Europe	60.6	70.2	43.8	31.5	24.6	31.5	32.1

Legend: *OHD - Occupational Health Department*
              HSCOMM - Health and Safety Committee

In Spain and Germany approximately 90% of companies had health and safety committees, whereas only 41% of Italian companies had these committees (legislative provisions explains much of these differences). Interestingly, more than two-thirds of the Irish and UK samples reported having Health and Safety committees, despite the fact that legislation does not dictate their presence.

The relative health priority reported by companies also varied between the countries, with Italy, Greece and Spain reporting higher levels of health priority than the other countries. Examination of the percentages of companies which gave health a low priority reveals that the highest prevalences were seen in the UK, Germany, Holland and Ireland. This finding may indicate that there is a strong north/south split in the priorities given to health by the companies. It is likely that cultural differences also have a large role to play in explaining these differences.

Only 31.5% of companies in the entire sample reported having written, explicit health policies. However 49% of Greek companies claimed to have such policies, while only 17% of Germans had. A similar pattern was seen in relation to the existence of health budgets, with 32.1% of companies in the entire sample having specific health budgets. In Greece, almost 48% of companies claimed to have health budgets, whereas as few as 23% in Ireland had specific health budgets. These findings are probably inflated by the Greek results, where the peculiarities of sampling may have led to respondents making inflated claims of their activities in the health area. (The survey in Greece was carried out by a team from the Labour Inspectorate).

In general, these results indicate quite large differences between the countries in relation to many of the demographic and health characteristics of the companies in the sample. However, with the possible exception of company size, there is sufficient variation within each of the national samples for analysis to take place on the European sample as a whole (appropriate statistical tests reveal this to be the case).

5.3 What was most important in explaining health activity ?

The results from the second stage of the analysis on the twelve indices of health activity indicate the contribution of company demographic factors, company health characteristics, the factors which prompted the health action and the personnel involved to explaining levels of health activity.

In general, the country variables were most important in explaining all types of health activity. The differing legislative backgrounds, the different levels of economic development, the different samples drawn in each of the countries and the differing traditions of the practice of health in the workplace go some way to explaining these national differences. However, there were some fascinating insights to be gained from examining how the factors which operate at company level - demography, health characteristics, the factors which prompt health actions and the personnel involved - combine to predict different types of health activity. The findings regarding these kinds of factors are of importance in framing recommendations targeted at company level.

5.3.1 What explains health screening activities ?

In general there was quite a good level of prediction of levels of health screening activity - 38% being explained in the case of the general index and 36% in the case of actions which had a health consideration.

In relation to the general index the strongest associations tended to be with country, with the UK, Greece, Germany and Ireland, all having lower levels of health screening activities than Spain. This is probably due to the fact that in Spain, health screening activities are compulsory for enterprises employing more than one hundred people.

However, other factors were also strongly associated with health screening activities taking place. The principal among these was the presence of an occupational health department, but they also included the presence of a health budget, presence of a written health policy and the existence of a health and safety committee.

The only demographic factor associated with health screening activities was company size, where larger companies tended to undertake more health screening activities. This finding is not surprising, as these are the kinds of company which tend to have more resources available for health activities of this kind. The only participation variable which predicted health screening activities was levels of involvement of health and safety representatives where higher levels of involvements tended to be associated with higher levels of screening activities. This finding is probably due to the involvement of health and safety representatives in the monitoring of screening of 'at-risk' individuals. None of the prompting factors for undertaking health activities were associated with health screening activities.

All of the above factors were also associated with health screening which had health improvement as a major consideration in their taking place, with the exception of the presence of the health and safety committee and the involvement of health and safety representatives. In addition, two of the prompting factors were associated with these kinds of health screening activities. These were the presence of health problems as a prompting factor and absenteeism. The significance of both of these factors makes intuitive sense - where there are higher levels of perceived health problems and where there are absenteeism problems, health screening provides a good starting point for understanding the nature of the problem.

5.3.2   What explains healthy behaviour activities ?

A significant number of factors were associated with the promotion of healthy behaviour activities. Levels of prediction were again quite high with 36% and 34% being explained in relation to the general and the health index respectively.

In relation to the first index, two demographic factors were important - the size of the organisation and the level of trade union membership, both of which were associated in a positive direction i.e. larger companies and companies which had higher levels of trade union membership were more likely to undertake these activities. While it is not surprising that the size of the organisation is important, as they tend to have more resources available, it is surprising that trade union membership was important as trade unions have often expressed suspicion of health actions which have been directed solely at the behaviour of the individual. (They fear 'victim blaming' and 'scapegoating' along with a fear that activities such as these are used as a substitute for actions aimed at the environmental causes of ill health).

There were also some national differences where in Spain, Italy and Greece lower levels of healthy behaviours activities took place than would be expected.

Company health characteristics were also associated with promotion of health behaviour activities. These included the presence of a health and safety committee, a health policy and an occupational health department. These findings are consistent with the commonly held perception that health actions are largely concerned with altering the individual's health behaviour. Though it is not surprising to see the factors of health policy and occupational health departments being associated with these activities, it is less intuitive that health and safety committees should be, given their traditional focus on the prevention of accidents, which is after all a combination of environmental intervention and the promotion of safe behaviour.

High levels of involvement by management and outside consultants were associated with promotion of healthy behaviour. These findings are consistent with perceptions of the marketplace, where many external consultants appear to rely solely on individually oriented interventions as a means of promoting health (see Hauss, 1991 for an analysis of the role of external consultants).

Four of the prompting factors were also associated with promoting healthy behaviour. These concerned legislation, staff morale, public image and accident rates. While it is easy to understand that improvements in staff morale and public image may be effected by behaviour modification programmes, it is less easy to understand why legislation and accident rates should prompt activities of this kind. Perhaps the explanation concerns the perception that any health activity is an appropriate response to problems in these areas.

Turning to health behaviour activities which had a health component, a somewhat different picture emerges where of the demographic factors, only size was associated with this index. In addition, Ireland, the UK and Germany tended to have higher levels of these activities taking place.

In relation to company health characteristics, the presence of a health and safety committee, health policy, and a health budget were associated with healthy behaviour activities which had a health component in them. The significance of a health budget in this context perhaps relates to the fact that external consultants (who provide many programmes in the are of healthy behaviour) tend to offer these kinds of programmes.

Of the personnel involvement factors the involvement of external consultants and occupational health staff was associated with this index, while of the prompting factors only legislation, staff morale and accident rates were associated with health behaviour activities. These findings are not surprising given the tendency of consultants and occupational health staff to offer these kinds of programmes.

5.3.3   What explains organisational interventions for health ?

It was more difficult to explain the levels of organisational activities when compared to the previous indices of health activity, with only 20% and 16% levels of prediction being achieved in relation to the general and the health indices. This relatively low level of prediction is perhaps understandable as the potential number of reasons for engaging is such activities is large, and they may legitimately be undertaken for reasons other than health. Furthermore, many of the interventions relate to activities which take place with a low level of frequency. It should also be noted that the independent factors used in the study were not selected specifically for their potential to predict these types of activity.

The only demographic factor of significance was the size of the enterprise, where larger organisations undertook more of this kind of activity. The country of origin was also of significance in relation to the general organisational interventions index. Levels of these interventions were higher in Spain and lower in the UK than in other countries.

The company health characteristics associated with the general index are health and safety committees, health policies, health budgets and having a high health priority. These findings are particularly interesting, since they indicate the potential of health and safety committees, backed up by appropriate resources, to insert the issue of health into ongoing company activities. (This being one of the goals of workplace health promotion). It is also notable that the presence of occupational health departments was not associated with organisational interventions, which is an indication that they either they are marginalised within companies, or that they do not see their role as being involved in such activity.

None of the participation factors were associated with these interventions, while the prompting factors of staff morale and accident rates were associated with the general index of organisational interventions.

Company size was the only demographic factor associated with organisational interventions which had a health component. In the UK these kinds of activities were least common. The presence of a health and safety committee was the only company health characteristic associated with organisational interventions which had a health component. Among the prompting factors associated were those of personnel problems, staff morale, absenteeism, staff turnover and public image, all of which were associate with organisational interventions with a high health component. These findings broadly support the findings on the general index. The fact that more prompting factors which are related to organisational performance were found to predict this index, suggests that companies see the potential of such activities to influence not alone organisational performance, but also the health status of their workers.

5.3.4 What explains safety and the physical work environment activities ?

Activities which focused on safety in a physical work environment also showed quite low levels of predictability - only 23% and 16% of the variance respectively was explained in relation to the general index and the index of activities which had a health component.

Looking at the general index first, none of the company demographic factors explained were significantly associated with them, though in Spain higher levels of these activities tended to be reported (this may be a result of the idiosyncrasies of the samples which were drawn from these countries).

The company health characteristics of health and safety committees, health budgets and occupational health departments were also associated with safety interventions. These findings are understandable as the kinds of activities which make up this category largely relate to traditional health and safety interventions.

Two of the participation factors were associated with the general index - the involvement of occupational health staff and that of health and safety representatives. This finding is agin very understandable in terms of the functions of these two types of personnel. (It would have been very strange to have no relationship between these groups and safety interventions !)

Four of the prompting factors were also associated safety and physical work environment activities. Not surprisingly legislation was associated with it as were staff morale, absenteeism and productivity/performance. These findings confirm the impression that safety activities are far better integrated into ongoing company operations than health activities with little or no safety component.

Safety interventions which had a health component in them were not well predicted by the factors used in the analysis. Only five factors were significantly associated with these kinds of interventions. These included the Netherlands, where lower levels of these activities took place, the presence of an Occupational Health Department and the prompting factors of staff morale, productivity/ performance and public image. These latter findings indicate, at least to some degree, that safety and physical work environment activities are not widely perceived to have a health component.

5.3.5   What explains social and welfare activities ?

In relation to social and welfare provisions, moderate levels of prediction were attained, with 28% of the general index being explained and 24% in the case of the health index.

Company size and levels of trade union membership were associated with both indices, such that larger companies with higher levels of trade union membership engaged in more of these activities. This finding is not surprising, as larger companies would have more resources to be active in this area, while trade unions have traditionally been active in seeking activity in the welfare area.

In Spain, the UK, Italy and the Netherlands fewer of these kinds of activities tended to take place while in Ireland more social and welfare activities tended to take place which had a health component in them.

In relation to company health characteristics, the presence of health policy, health budgets and occupational health departments were all associated with higher levels of these activities in relation to both the general index and the health index. Again thee findings are not surprising as welfare activities have often been associated with the occupational health function, while the expression of health policy often has a welfare component (see Johnson & Johnson's 'Credo' for example).

In relation to prompting factors, staff morale and accident rates were associated with the general index as they were with the health index, while personnel problems and absenteeism were also associated with the health index. Again none of these factors are surprising with the possible exception of accident rates. In this case accident rates may prompt companies into rehabilitation programmes which come under the heading of welfare.

5.3.6 What explains overall levels of health activity

The overall indices of general activities and activities which had a health component were predicted best of all by the combination of factors used in the analysis, where 40% and 29% levels of prediction were attained. (These indices comprise the sum of all of the activities examined).

Significantly none of the countries were associated with higher or lower levels of these activities, despite the fact that many were associated with the more specific health activities.

Company size was associated with both indices, while levels of trade union membership was associated with higher levels of activity in the general index. These findings are not surprising as company size was consistently associated with the more specific indices. The failure of the factor of levels of trade union membership to predict the health index is somewhat surprising, and may indicate that trade unions have not been as active in the area of health activity as they profess to be. Additionally, it may indicate that trade unions are not aware of the health improvement potential of many of these activities.

The presence of a health and safety committee, a health policy, a health budget, having a high health priority and an occupational health department were all associated with the general index, while only a health and safety committee and a health budget were associated with the health index. These findings indicate that the groups who are most active on the shopfloor tend to be most aware of the potential for health improvement and that they are the agents which can most effectively integrate health issues into company activity. The failure of the presence of occupational health departments to do likewise raises serious questions about their role in companies - are they marginalised by company practice or are they unwilling to adopt the function of integrating health concerns into company practice (or both) ?

The involvement of occupational health staff was associated with the general index but significantly, it was not associated with activities which had a health component in their taking place. This finding supports the view outlined in the previous paragraph that the role of occupational health and safety personnel needs examination.

The involvement of outside consultants tended to be associated with the health index but not with the general index. This finding is not surprising, as presumably external consultants carry a specific health brief. It is somewhat surprising to find that their involvement predicts health activities at all in the light of the failure of this factor to regularly predict the more specific indices.

Four of the prompting factors were associated with overall activity. These were personnel problems, staff morale, productivity/performance and public image, all of which were associated with activities which take place for health reasons. Of the prompting factors, only staff morale and public image were associated with general levels of activity in this area. It is interesting to note that problems with personnel or productivity/ performance were associated with health activities, as this finding argues for some level of effective integration of health issues into ongoing company practice. Even if this integration does not occur in a systematic way, the finding suggests that health activities are seen by enterprises as a possible means of addressing problems in these areas.

5.4   So what <u>does</u> influence workplace health actions ?

The findings from these analyses provide a complex and rich picture of the factors which are uniquely associated with health activities and the involvement of the major workplace actors in these activities.

It is clear that the principal predictors of levels of health activity resided perhaps surprisingly not with the country from which the data emanated, but with the health characteristics of the companies concerned. For example, the presence of a health budget and of an occupational health department predicted nine of the twelve indices of health activity while the presence of a health policy and of health and safety committees predicted eight of the twelve health activity indices.

These findings have clear implications for the establishment of health actions in the workplace in so far as the presence of these company health characteristics actively aid the establishment of health activity and health promotion in the workplace. In particular, the presence of health and safety committees appears to ensure at least some level of effective integration of health issues into company practice. (See Recommendations for further information).

Of the demographic factors, only size and the level of trade union membership were associated with the health activity factors. Size was very strongly associated in most cases, in fact in relation to ten of the twelve indices of health activity. The level of trade union membership however was much more weakly associated - in this case with only four of the twelve health activity factors. The findings in relation to the company size are not surprising where it is the received wisdom that smaller companies tend to engage in fewer health activities.

However it is somewhat surprising that levels of trade union membership were not more strongly associated with high levels of activity. This failure to find significant associations with levels of trade union membership begs the question of the role which trade unions actually play in the establishment of health activities in the workplace. It would appear that there is considerable room for improvement in this regard.

In relation to the different countries, the findings are again not perhaps surprising. The UK and the Netherlands were associated with six out the twelve health activity factors each. In these cases in the Netherlands tended to produce higher levels of activity, the UK lower. Interestingly the data from the Irish sample tends to match the levels seen in the UK sample.

The prompting factors also had a role to play. The issue of staff morale was particularly strongly associated - in this case with ten of the twelve health activities. This finding would indicate that where staff morale is given as a reason for the establishment of health activities, then health activities are far more likely to take place. Interestingly the presence of health problems were associated with only one of the health activity factors. Other important prompting factors concerned public image, accident rates and absenteeism rates, which were associated with five, five and four of the health activity indices, respectively. While these findings are complex, it would appear that where activities are prompted by factors other than health problems, there is a greater likelihood of them taking place.

The levels of involvement of the six main actor groupings tended not to be associated strongly with levels of health activity. The only exception to this rule was the involvement of occupational health staff, which was associated with five of the twelve health activity factors. Interestingly, the involvement of staff representatives and of union representatives was not associated at all with levels of health activity. This again begs the question of the role of trade unions in the establishment of health actions in the workplace.

6. Establishing Health Actions

This chapter looks at the ways in which organisations establish health actions, at the factors which prompt organisations to undertake health action in the first place, at the kinds of personnel involved and in particular at some of the difficulties faced by organisations in establishing workplace health action. It deals with the barriers to workplace health actions and also with the benefits to be gained from them. Successful approaches to overcoming the commonly occurring problems facing organisations are outlined.

6.1 Why do organisations undertake health actions ?

The predominant factors which prompted health actions were legislation and morale in the workforce (see Figure 6.1). These were closely followed by personnel problems, health problems, productivity problems and levels of absenteeism. Factors relating to labour turnover and industrial relations problems were not often cited as reasons for engaging in health actions in the workplace. This pattern of findings indicates that organisations react to both external pressure and internal demands when undertaking workplace health action, and that there are a multiplicity of reasons for them undertaking action.

Organisations also experience benefits from engaging in health actions - whereas they react to the push factor of legislation, there is also a pull factor of benefits to be gained. Examination of the benefits that accrue from engaging in health actions (see Figure 6.2) reveals that the principal benefits were found to be in terms of morale and health improvements of the workforce. In addition, significant benefits were claimed by more than 50% of the companies in terms of reduced personnel and welfare problems, reduced absenteeism, increased productivity, reduced industrial relation problems, improved company image, and reduced accident rates.

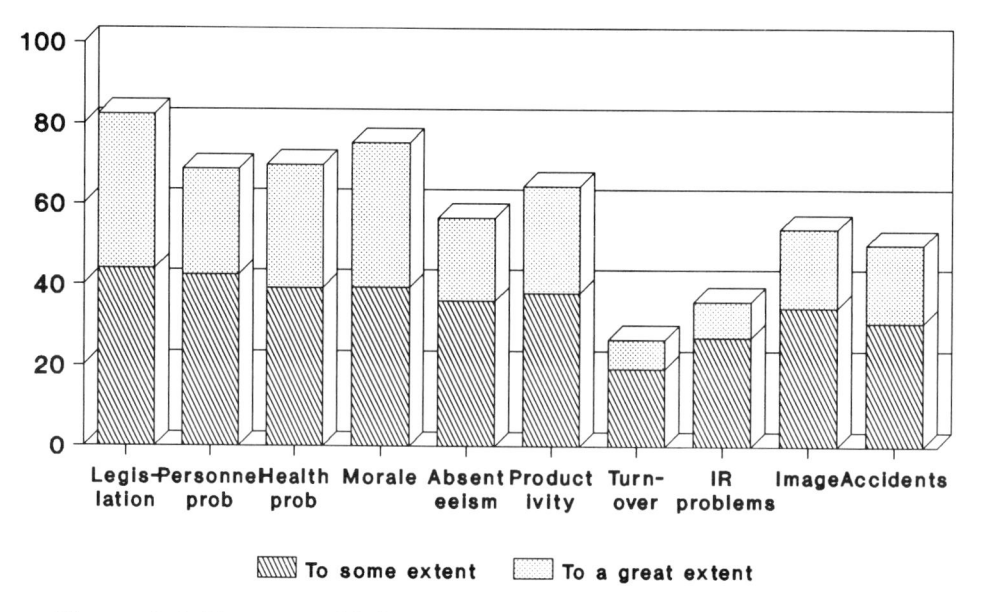

Figure 6.1 Factors which prompted health actions

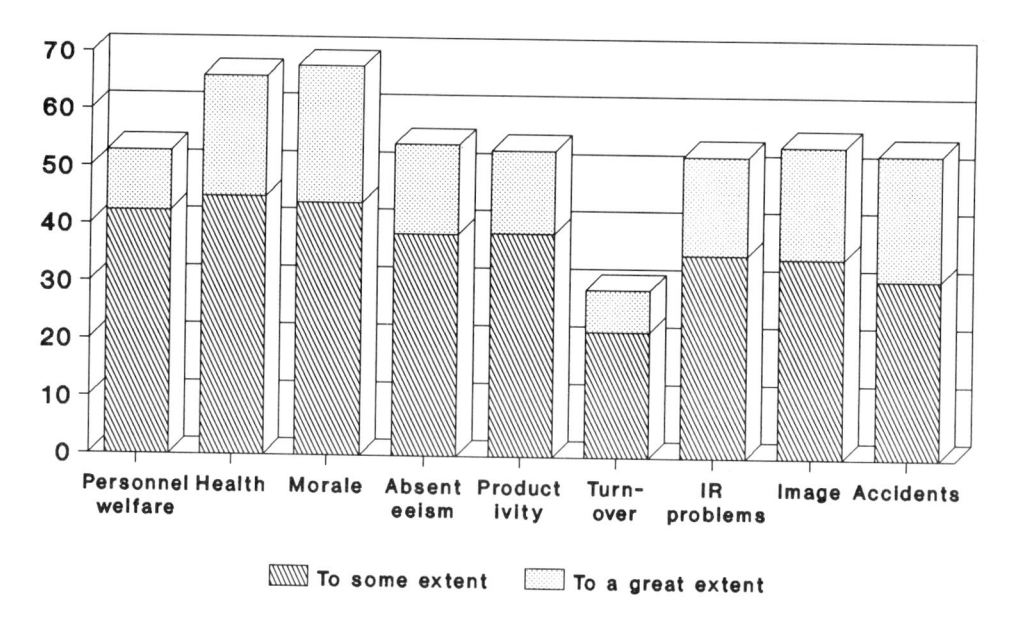

Figure 6.2 The benefits of health actions

It is interesting to contrast the reasons for undertaking health actions with the benefits which companies cited as taking place. Table 6.1 illustrates these findings. In this Table what can loosely be termed an awareness gap or index can be calculated, where the percentage of respondents who report an issue as prompting a health action is subtracted from the percentage who reported seeing benefits in that particular area.

This leads to some interesting findings, particularly in relation to areas such as industrial relations. Of all of the prompting factors, the second least important (industrial relations problems) yielded considerable benefits as a result of health activities. The awareness index of 20.3 indicates that the despite the apparent benefits of health actions on industrial relations climate, problems in this area do not appear to prompt health actions to a commensurate extent. Conversely, while industrial relations problems are relatively rarely cited as a reason for engaging in health action, considerable benefits occur in this area. Other issues where an excess of benefits were seen included staff morale, staff turnover, accidents rates, and company image. For most of the rest of the areas, the percentage of respondents reporting benefits was almost the same as those who reported the issue as being a prompting factor, with the exception of the personnel and welfare area and the productivity/performance area.

Table 6.1 Prompting factors and the benefits of health actions

	% Reporting as a prompt (1)	% reporting as a benefit (2)	Awareness gap (2-1)
Personnel/Welfare	76.3	62.4	-13.9
Health	77.3	75.8	-1.5
Staff Morale	72.7	77.8	+5.1
Absenteeism	63.3	62.8	-0.5
Productivity/Performance	71.7	62.3	-9.4
Staff Turnover	29.3	35.5	+6.2
Industrial Relations	41.3	61.6	+20.3
Company Public Image	60.8	64.2	+3.4
Accident Rates	56.2	63.7	+7.5

These figures should not be interpreted too rigidly. What they essentially show is that companies have multiple reasons for implementing health actions, many of which have nothing to do with health issues and secondly that large proportions of companies indicated that they have perceived benefits from implementing health actions. It would appear that in relation to the benefits in the areas of industrial relations and to a lesser extent accident rates and staff turnover that companies are comparatively unaware that benefits would accrue in these areas as a result of implementing health actions.

Respondents were asked to nominate the most important benefits they perceived as a result of engaging in health actions. The three most important findings from these open-ended questions are outlined in Table 6.2.

The most notable feature of these findings is that relatively large numbers of companies spontaneously reported benefits, and that these benefits were to be found in both the harder measures such as reduced absenteeism and increased productivity and in softer terms such as improved social climate or staff morale, improved company image and improved employee motivation.

It was also striking that only one of the countries (Germany) spontaneously reported health improvements as one of the three most important benefits. This finding could mean that organisations are unaware of health benefits that accrue or that these benefits do not happen with a high frequency. In either case their is a clear need to adequately demonstrate the health benefits which can result from well designed health programmes.

Table 6.2 Spontaneously reported benefits of establishing health actions

Country	Benefits
Germany	Health improvements Improved social climate Improved company image
Greece	Productivity Accident reduction Improvements in industrial relations
Ireland	Staff morale Health Absenteeism/productivity
Italy	Reduced accidents and disease Improved industrial relations Improved work environment
The Netherlands	Reduced absenteeism Improved ergonomics Improved employee motivation
Spain	Improved industrial relations Decreased absenteeism Improved health

6.1.1   Multinationals, health policy and motivation

Evidence from the case studies of the multinationals sheds light on why they have engaged in health actions. In many cases the reasons given relate to company culture, where overt statements of human resource management policy may include maintaining and protecting the health of the workforce. US firms in particular have tended to adopt this approach. These policy statements can confine themselves to a search for excellence in all that the company undertakes, and this search extends into production and human resource management areas. An example of such a policy is provided by a US multinational, where inter alia, the following is stated:

> "...... *We are responsible to our employees, the men and women who work with us throughout the world. Everyone must be considered as an individual. We must respect their dignity and recognise their merit. They must have a sense of security in their jobs. Compensation must be fair and adequate, and working conditions clean, orderly and safe. We must be mindful of ways to help our employees with their family responsibilities ...... There must be equal opportunity for employment, development and advancement for those qualified. We must provide competent management and their actions must be just and ethical ......*"

This company has used this statement of company philosophy as a basis for developing one of the largest health programmes existing.

In addition to these human resource management policies, many of the multinationals reported having overt health policies or philosophies. One UK based subsidiary of a German multinational reported the following aim of occupational health policy:

> "*To provide a comprehensive Occupational Health Service within the Company UK Health and Safety Policy, to promote and maintain the highest possible degree of physical and mental health of staff in so far as is reasonably practical*"

However, there can be confusion regarding the exact meaning of such statements as the company also claims that they do not have a written health philosophy !

Health policies may have more venial purposes. An example from Spain (OPQ - see Box 2 below) illustrates how health policies can have an ulterior motive which can lead to resistance and hostility. The health policy of OPQ is based on the belief that "a healthy worker is a good worker", and in practice this has meant that the application of this policy has been to produce good workers rather than healthy workers.

The dangers of employers using health activities as a means of controlling the workforce are real and apparent to many workers. The use of medical checks as a basis for selection, redeployment and redundancy practices appears to be quite widespread, and has lead to a common perception that occupational health professionals are tools of management.

A further tendency was evident in many companies which use dangerous substances in their production processes - in these companies there tended to be an appropriate level of concern for safety, often far in excess of the legislative provisions of the various countries they operated in. This concern with safety then 'spilled over' into a concern for the health of the workforce. This motivation for engaging in health activities was seen in two companies in particular, operating in the petrochemicals and computer manufacturing sectors. Further evidence from the multinationals would suggest that the improvement of company image had a role to play in developing their health policies and practices.

Though much of the evidence from the multinationals suggest a range of 'rational' reasons for their health activities, the role of motivated and powerful personnel in key positions should not be underestimated. In many of the companies examined, the presence of energetic and visionary members of staff who have grasped the opportunities available for health activity was notable. These individuals were located at many points throughout the company, ranging from head office to plant level. It was notable that even in companies with strong overt health policies that the implementation of these at plant level often depended on the presence of motivated and committed individuals.

A detailed discussion of the findings from the multinational case studies in relation to health policy is to be found in Hauss (1992).

6.2 Who participates in these health actions ?

Figures 6.3 and 6.4 summarise some of the results from the section of the questionnaire which dealt with levels of involvement in health actions by six major actor groupings. These are management, staff representatives, trade union representatives, health and safety representatives, occupational health staff and external consultants.

Two specific issues were looked at in relation to levels of involvement. The first issue concerned an attempt to chart the different levels of involvement of these actors across what might be termed the life cycle of a health action. The life cycle of a health action was conceptualised in terms of four stages involving the initial idea for the health action, the planning of the health action, the implementation of the health action and the evaluation of the health action.

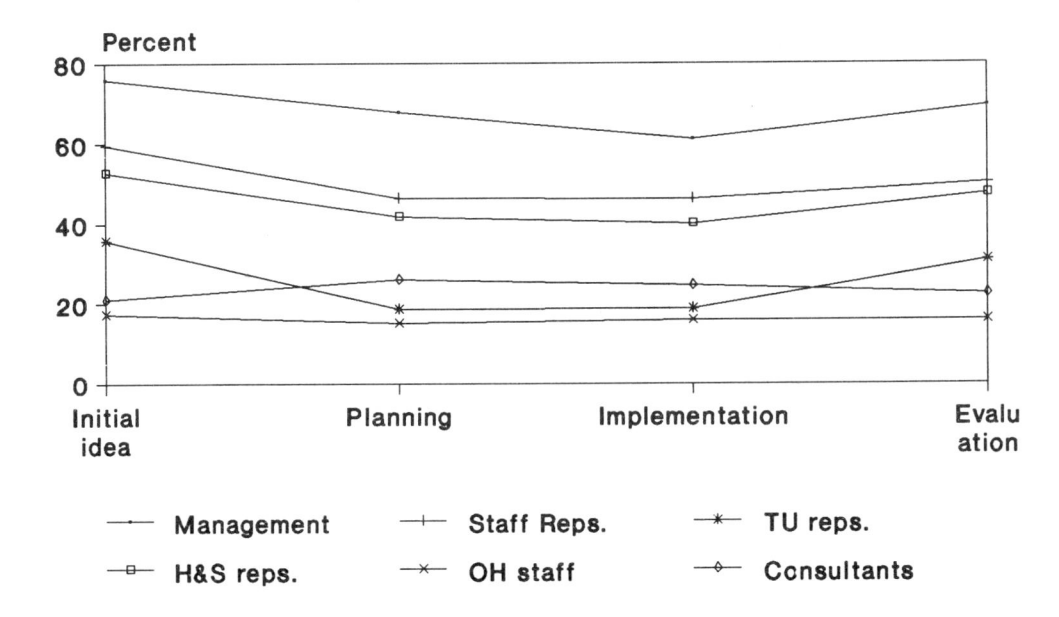

Figure 6.3 Participation during the life cycle of health actions

The second issue which was examined concerns the level of involvement of these actors during the implementation stage of the health action life cycle. Four levels of involvement were hypothesised. These involved at a minimum, the provision of information, followed by consultation (which involves views of the workforce being sought, but not necessarily being integrated into decision making), participation (where there is free information flow, free consultation, but the workforce have an equal input into decision making) and finally responsibility (where all three previous levels of involvement are satisfied but in addition the workforce not alone participate, but are also responsible for implementing the consequences of decisions made).

Looking at levels of participation during the life cycle first, it is apparent that management had by far the highest levels of involvement, with between 60% and 80% of managers reporting involvement at all stages of the life cycle. These are followed by staff representatives and health and safety representatives. Following these, with quite low levels of involvement were trade union representatives, outside consultants and occupational health staff.

It is also of interest that most groups exhibited a particular pattern of involvement, where the highest levels of involvement tended to be seen during the initial idea stage, and levels of involvement fell during the planning and implementation stages and rose slightly again during evaluation. Different patterns were seen for external consultants and occupational health staff, where they tended to maintain their level of involvement throughout the life cycle of a health action. This would seem to indicate that professional or technical staff are involved in a more constant way while non professional staff are involved in setting policy on the one hand and evaluating effects of implementation on the other.

Figure 6.4 illustrates levels of involvement of the six groups during the implementation stage. Management had the highest forms of involvement, while all other groups had lower forms of involvement, particularly in relation to the distinction between participation and responsibility.

Interestingly, the lowest levels of involvement were reported for the occupational health staff, external consultants and trade union representatives. It appears that while technical staff tend to maintain a constant involvement throughout the life cycle of an activity, the type of involvement which characterises their activities tends to be quite low.

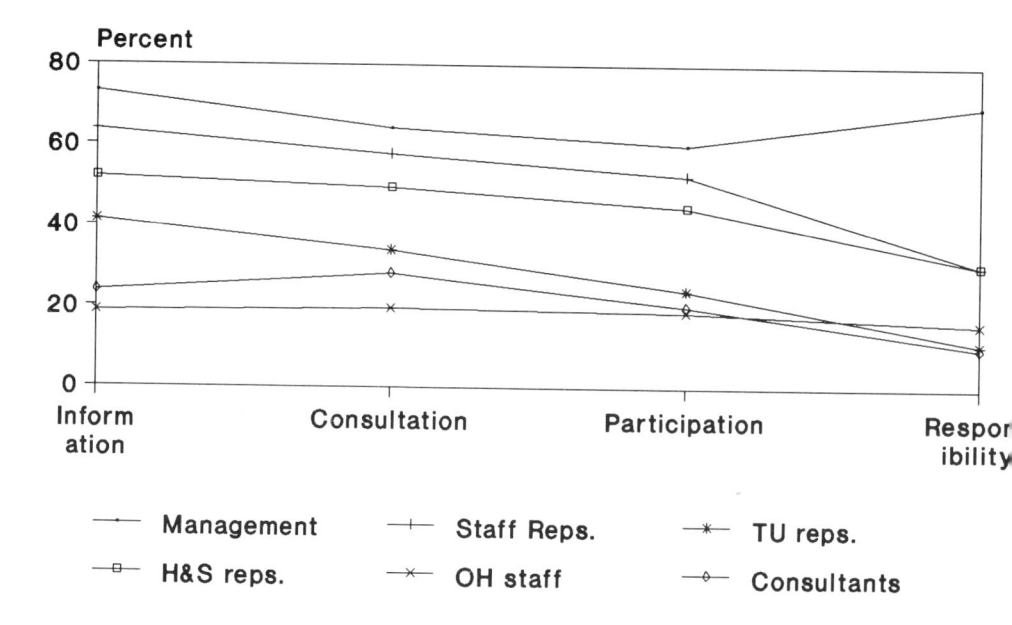

Figure 6.4 Participation during the implementation stage

The evidence from the multinational case studies provides a rather richer picture of patterns of involvement in health actions. In countries such as Germany, where rights to participation are enshrined in legislation and traditional practice, participation in health activities tended to follow the route of these traditional structures (e.g works councils). In other countries such as Ireland and the UK where participation in decision making by workers has not been so formalised, the type of involvement was also less formalised. Typically, workers were informed and consulted about health activity plans, but often in a manner which had no formal status. Among the more active companies this style of participation seemed to work well, but it was notable that less active companies tended to be characterised by very low levels of even informal participation.

6.3 Does involvement predict health activity ?

A specific analysis of the relationships between levels of participation and the health activity indices was performed. This analysis was of particular interest in order to closely examine the effects of participation on health activity, as it can be hypothesised that high levels of participation, particularly by employees would be associated with more successful health actions, and by implication with higher levels of activity.

Table 6.3 outlines the most important predictors for each of the twelve indices of health activities. Perhaps the most remarkable aspect of these findings is that participation levels were not very strongly associated with any of the health activity variables, with levels of prediction ranging from 6% to 21%. It is perhaps also surprising that the amount of variance explained was less in the case of health activities which had a health consideration.

Table 6.3 Most important predictors of health activity

	Consul-tants	Union Reps	OHS staff	Manage-ment	H&S reps	Staff reps
Screening 1	ns	ns	+	ns	+	ns
Screening 2	ns	+	+	ns	+	ns
Healthy Behaviours 1	+	ns	+	+	+	ns
Healthy Behaviours 2	+	ns	+	ns	+	ns
Organ Interventions 1	+	+	ns	+	+	ns
Organ Interventions 2	+	ns	+	+	+	ns
Safety and environment 1	+	+	+	+	ns	+
Safety and environment 2	+	+	+	ns	ns	+
Social and Welfare 1	+	ns	+	ns	ns	ns
Social and Welfare 2	+	ns	+	ns	ns	ns
Total 1	+	+	ns	+	+	ns
Total 2	+	ns	+	ns	+	ns

Of the participation variables, involvement by Occupational Health Staff and Health & Safety representatives tended to be most strongly associated with health activities. Also of interest was the finding that levels of involvement by outside consultants was consistently, if weakly, associated with all of the health activities with the exception of screening activities. Despite having by far the highest levels of involvement, participation by management was not consistently associated with health activity. Of particular interest here was the tendency for management involvement not to predict activities which had a health component in their establishment.

6.4 What are the barriers to health action ?

The fact that the organisations which responded to the questionnaire were precisely those companies who have overcome many of the barriers to establishing workplace health action means that important insights can be gained from the survey regarding the problems that they did face. In addition, the case studies of the multinational companies also provide valuable lessons regarding practical approaches to overcoming problems (and to creating them !).

Information addressing the issue of barriers and strategies to overcome barriers comes from a number of sources. On the questionnaire instrument two approaches were taken to the issue - the first came from fixed choice questions regarding the extent to which a range of issues acted as barriers, while the second came from open ended questions regarding the nature of these barriers. In addition, further questions were asked regarding the strategies used to overcome specific barriers. The information collected on the benefits of workplace health action are relevant, as they deal with many of the factors which had already been cited as barriers. The information collected as part of the multinational case studies is also of relevance in this regard.

Respondents to the survey were asked to indicate the kinds of problems that they experienced in relation to establishing health actions in the fixed response section of the questionnaire. Of these the most significant were financial resources, lack of suitable facilities, what would be perceived to be a lack of staff commitment, closely followed by a lack of expertise, and a lack of human resources. Interestingly, the least cited problem was lack of management commitment, though this finding is possibly a function of the kinds of people who actually completed the questionnaire.

The findings from the open ended part of the questionnaire generally supported these findings. Table 6.4 indicates the principal problems which were spontaneously reported in each of the countries for which data was provided.

Table 6.4 Spontaneously reported problems in establishing health actions

Country	Problem
Germany	Lack of commitment
	Technical/space problems
	Finance
Greece	Finance
	Employees attitudes
	Lack of specialist staff
Ireland	Finance
	Management commitment
	Employee commitment
Italy	Finance
	Achieving consensus
	Employee commitment
The Netherlands	Finance
	Lack of manpower/time
	Employee commitment
	Management commitment
Spain	Employees attitudes
	Finance
	Company structure/organisation

Respondents were also asked to indicate the kinds of problems that they experienced in relation to establishing health actions on fixed-response format questions (see Figure 6.5). Of these the most significant were financial resources, lack of suitable facilities, what would be perceived to be a lack of staff commitment, closely followed by a lack of expertise, and a lack of human resources. Interestingly, the least cited problem was lack of management commitment.

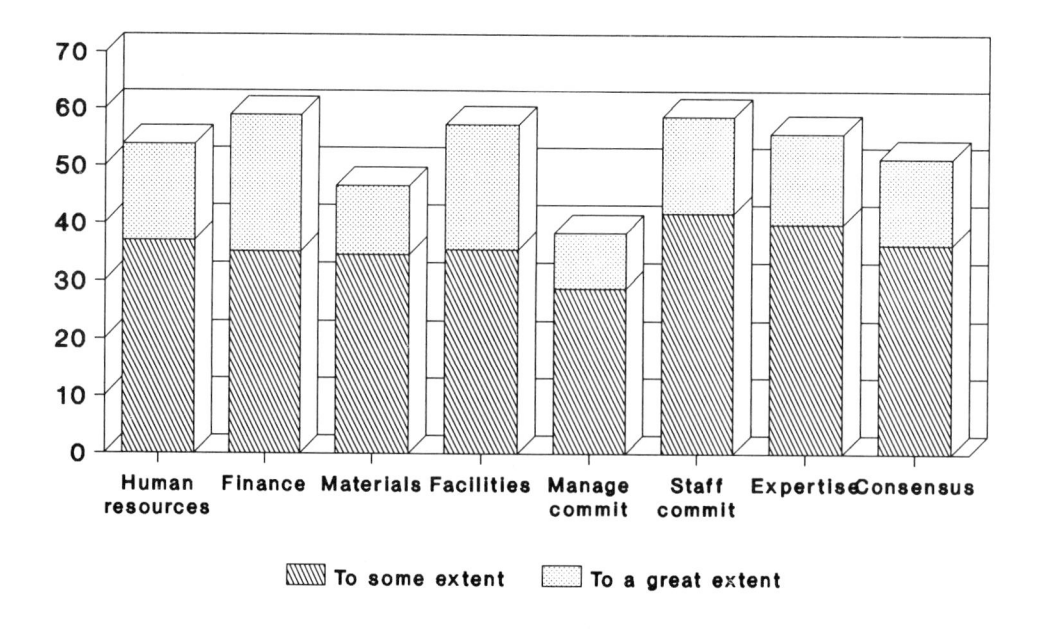

Figure 6.5 Problems in establishing health actions

This last finding is almost certainly a function of those who actually completed the questionnaire. Table 6.5 illustrates the job titles of the individuals who completed the questionnaire and it can be seen that there was a disproportionate number of management, who might be more likely to cite staff commitment as a problem as compared to staff who might be more likely to cite management commitment as a problem. What is perhaps nearer the truth are the numbers who report difficulties in achieving a consensus between the key actors in the workplace as a problem. The prevalence of this problem is almost exactly half way between the numbers reporting lack of management and lack of staff commitment.

Table 6.5 Job titles of respondents who completed the questionnaire

Job Title	%
CEOs	20.9
Personnel	36.6
Occupational health professional	10.2
Health and safety representative	4.4
Finance	3.8
Other	24.1

6.5 How are barriers to health action overcome ?

Companies were also asked to indicate the successful strategies which they had used to overcome some of the most prevalent barriers. Generally speaking, the numbers of responses to this question were low and it is difficult to generalise from the findings. However, it is apparent that many of the most prevalent problems can be solved at least in a small number of respondent companies. Table 6.6 indicates some of the strategies used to over come problems. This Table is based on accounts supplied in each of the national reports, though not all of the national reports dealt with this issue.

One of the striking features of Table 6.3 is that most of the strategies reported were in relation to improving employee or management commitment to the issue of health promotion (this being one of the most prevalent problems reported). Quite a wide range of strategies were reported in this regard. These included making senior management commitment visible and explicit, setting up structures within the workplace which have specific responsibilities for the health promotion action (e.g. staff action groups, designating specific health promotion roles) and including the issue of health promotion in formal management-labour contacts (e.g. works councils) within the workplace.

In relation to finance, the planning of actions was mentioned as a successful strategy particularly on the grounds that proper plans with proper costings attached to them tend to convince management of the value of undertaking health actions. In addition, the actual provision of finance and resources for the health action overcomes this problem. While this strategy might be viewed as being simplistic, it should be seen in the light of the benefits to be gained from the health action which would appear at least in some managements to outweigh the costs of implementing it.

Table 6.6 Successful strategies used to overcome barriers to health actions

Problem	Strategy
Employee/management commitment	Information/training
	Payments to employees
	Support of senior management
	Emphasising legislation
	Inclusion of issue in formal talks between management and workers
	Making health actions compulsory
	Competitions
	Giving feedback
	Improving resources for health promotion
	Creating staff action groups
	Accurately defining the magnitude of the health issue
	Designation of specific roles
Finance	Planning of actions
	Co-operation with public services
	Creating a budget
	Improving resource allocation
	Demonstrating management commitment
Facilities	Confining actions to specific settings
	Planning/building facilities
Lack of experience	Designation of specific roles
Lack of manpower	Additional staff
	Setting up a health promotion group

Fewer strategies are mentioned in relation to dealing with lack of facilities, lack of experience or lack of manpower, but again they tended to emphasise practical solutions to the problems, though these may in many cases cost relatively large amounts of money.

It is striking that in Table 6.2 that with the exception of Germany, productivity increases and reductions in accidents and improvements in absenteeism, all of which have concrete and measurable economic benefits to companies have been cited as being among the top three benefits to be gained from undertaking health actions. It is also arguable that in Germany for example, the health improvements and the improved social climate which were cited as benefits of health actions also have financial benefits to the company even if they are not as easily measured as some of the other benefits.

The implications of these findings are clear. If management is to increase commitment to health promotion in the workplace they need to be convinced of these benefits and also to be presented with comprehensive plans.

Some of the best models for planning health actions in the workplace are provided by some of the multinational case studies. A specific example of this is to be found in one of the Irish case studies. Further examples are to be found in the German case studies, outlined in Chapter 2. Box 1 contains an account of the planning activities of the multinational based in Ireland, which is an illustration of how planning and control can be achieved.

Box 1 - Multinational Case Study Report

Background

This case study refers only to the major Irish manufacturing plant of a computer multinational which employs approximately 780 people. The company employ a team of two occupational health nurses who provide occupational health services to the workforce at the manufacturing plant. In addition, a part-time Physician is employed by the company on a sessional basis. Occupational Health Nurses were employed by the company because health is seen as being an intrinsic part of company policy.

Box 1 Continued - Multinational Case Study Report

The company has a written health philosophy and a written health policy which is valid for all plants within all countries in which the group operates. This health policy appears to be fulfilled to the same extent in all of these countries and plants. The core of this philosophy as it is worked out within the Irish plant is that it focuses on environmental health and safety, preventive health care and health promotion. Other aspects covered by the health policy include the provision of direct care, continual professional education and employee assistance programmes.

<u>The Organisation of Health Promotion Activities</u>

Responsibility for Occupational Health and Safety and Health Promotion is located at plant level. However the parent company provide guidelines, leadership and materials to further activity in these areas. Headquarters also provides support services for medical information and there are consultant services available within company headquarters. Standards are established in consultation with headquarters for physical and psychological health which are to be implemented on a company wide basis. While there is no formal reporting relationship between the plant and headquarters, the parent company perform a Health and Safety and Environmental audit of each of the subsidiary plants every two or three years. The purpose of these audits is to support the subsidiary plant in its health and safety activities.

Decision making regarding safety practices are more centrally controlled in some cases (e.g. CFC reduction), with guidelines being issued from headquarters and adherence to these guidelines at a local level. In the majority of cases, however, safety practices are developed on the basis of hazard identification and risk assessment.

The decision making processes which take place regarding health and health promotion activities are conducted at a plant level. In the current case, a one year plan is drawn up annually which details the activities which are planned for the year and are agreed with management. In drawing up this annual plan changes are incorporated on the basis of reviews carried out by the Occupational Health team and by the personnel group within the company. When the plan is finally available it is communicated to all employees. Dates of implementation are fixed and assessment of findings from the activities are made available to all employees. In addition, individual feedback is a feature of many of these activities.

Box 1 continued - Multinational Case Study Report

The company does more than is legally required in both the health and safety arenas. For example, the Occupational Health team have insisted upon receiving safety data on all potentially hazardous substances used in the manufacturing process in advance of the substance being purchased so that risks are assessed and appropriate recommendations can be made regarding handling use and storage of the substance. These are provided by the suppliers of these substances and are issued to all employees either working with or exposed to these substances. In addition, employees in these categories receive chemical hazards awareness training. The relevant safety precautions indicated on these data sheets are actively incorporated into the handling of these substances by the workforce in everyday practice.

There is also certain amount of central health policy planning within the company. This planning particularly relates to safety issues where data on new hazards is gathered centrally at headquarters and guidelines for dealing with these new hazards are issued throughout the subsidiaries. These centralized planning structures do not receive input from subsidiaries in this process. This function is the responsibility of headquarters' health department.

Health promotion targets are set for the company as a whole and currently these focus on the prevention of occupational illness and injury and the promotion of wellness. In addition there are targets for the reduction of risks from chemicals which are used in the manufacturing process. Targets have also been set in relation to non-occupational illness, for example coronary heart disease and cancer.

There are a wide range of informal connections between the plant and other plants in relation to health promotion activities (much of the contact which takes place through these networks is conducted via electronic mail). They have strong connections to a sister plant located in Ireland, with whom ideas and advice are freely exchanged. Similarly, a lot of networking takes place with sister plants in other countries. The kinds of issues which are covered are of a general nature, but also includes specific advice on given activities. These contacts have arisen generally on the basis of personal contact.

<u>Participation</u>

Employee representatives are not formally involved in the decision making process on the planning and implementation of health actions. They are nonetheless regularly consulted on these plans. This occurs in an informal though seemingly comprehensive way, as the occupational health nurses take great care to sound out opinion through factory walk-throughs and, more formally, through regular short inputs to operational group meetings at the shop floor level.

Box 1 continued - Multinational Case Study Report

Financing

Financing for health activities in the workplace are handled through annual budget negotiations in which the plans for the year are put forward and budgets are assigned for them following a negotiation process. Once budgets have been decided there is full autonomy in how these budgets are to be spent for the occupational health department. If extraordinary issues arise during the course of a given year additional budgets can be sought and can be granted. The budget that is agreed upon is basically applied to ancillary or additional activities and does not include salary or overhead costs which exist in keeping the health department going.

By way of contrast to the above example of planning, Box 2 illustrates how planning should not be undertaken in an account of a multinational's activities in Spain. The way in which this company operates in Spain is completely at variance with how it manages it's activities in it's home country, where health activities are planned appropriately. In Spain, however, despite the appearance of a planning and implementation mechanism, a minimalist approach to both the planning and implementation has led to sever problems of uptake.

Box 2. Multinational Case study report (OPQ)

Organization of the enterprise

OPQ is a European multinational specializing in the manufacture and sale of electrical and electronic equipment, with establishments in several European countries. Head office makes decisions concerning OPQ's work units in Spain, and particularly those relating to personnel (including health issues). The central personnel office has power over factory personnel officers and other staff responsible for personnel at the various work units in Spain. OPQ has some 3800 employees in Spain, working in various units specializing in production, distribution and research. There are links with the parent company, usually through head office: the parent company monitors and receives reports on the projects and activities implemented at each establishment. There are also meetings of management representatives from each country.

> **Box 2 continued. Multinational Case study report (OPQ)**

Health policy, like any other issue coming under the jurisdiction of the personnel department, must be discussed in advance with head office. OPQ has a very rigid vertical hierarchical structure. It is an enterprise with a high rate of union membership among its workers and with a history of difficult industrial relations. The workforce has on several occasions taken strong industrial action concerning health and safety issues - action that has often hit the headlines.

<u>Health Policy</u>

The enterprise's health policy is based on the belief that "a healthy worker is a good worker". Good health is vital to good performance; a sick person is of no value, since he or she cannot, by definition, be a good worker. OPQ's ultimate objective is that its employees should work harder and better, and the enterprise acknowledges that growth in production is one of the prime factors behind its implementation of health actions. It also wants to achieve an improvement in industrial relations and a reduction in absenteeism. The enterprise claims to have a written health policy, though it is little more than a personnel policy. Management's definition of health is somewhat incomplete, since it is essentially limited to physical health, and this is reflected in the health actions implemented in the enterprise: medical checks only. No mention was made at any point of the mental and social aspects of health (for example, stress). OPQ retains the traditional definition of health as the absence of physical ailments. OPQ also seeks to comply with minimum statutory requirements concerning health and the quality of working life.

<u>Health actions</u>

According to management, OPQ's employees have no health problems, and in particular, none that are related to work. Indeed, if any health problem does arise, it is invariably of a common kind (colds, high cholesterol, etc). With respect to production workers, mention was made only of light industrial injuries such as sprains, problems caused by physical exertion, etc. There had been no serious industrial accidents. Workers may also have to handle paints, which may involve some risk, though the appropriate preventive measures are taken. There is a major lack of communication among the various workplaces for the exchange of information on health problems, personnel issues, etc.

> **Box 2 continued. Multinational Case study report (OPQ)**
>
> *Despite this official picture, OPQ has made a name for itself as the enterprise whose employees took part in one of the very few strikes Spain has ever witnessed in connection with health and safety at work - the continued exposure of a large number of its workers to noise well in excess of recommended levels. Given the dominance of physical health in the company's thinking, it is not surprising that health actions are limited to compliance with current regulations on health and safety and the conducting of annual medical checks. Concern for employees' health is equated with occasional medical examinations. The company medical service merely performs the traditional function of providing medical care, though slight progress is reflected in the gradual introduction of additional measures.*
>
> *Apart from conducting annual medical checks, the company medical service provides medical care and advice, as well as first-aid treatment in the event of minor injuries. The company medical service does not take any form of action to monitor or prevent problems not specifically related to the working environment, nor are there any plans for it to do so in the future. Occasionally, and in collaboration with the Mutua Patronal de Accidentes de Trabajo (body insuring employers against the risk of industrial accidents), it runs courses on first aid, etc. Perhaps not surprisingly, the enterprise has no special budget for health and safety or health actions; funds are allocated in accordance with the decision of national head office.*

The final illustration of how workplace health promotion is organised comes from Health and Welfare Canada (HWC), where the Federal Government has supported the development of a comprehensive method (Workplace Health System) for establishing health activity in a range of workplace settings. There are a number of unique features to the HWC model - it can establish activity in areas where there was none before, it is applicable to enterprises of all sizes, it works with but is not dependent on in-house services, it is under the control of the enterprises themselves (not controlled by external consultants) and is highly participative in nature. As a model it provides considerable possibilities for European workplaces to become engaged in workplace health promotion in a meaningful way.

Box 3. The Workplace Health System - Health and Welfare Canada

To keep the workplace healthier, from the office to the plant floor, Health and Welfare Canada has developed a comprehensive package designed to help Canadian companies put health programs and policies in place. It's called the Workplace Health System (WHS).

<u>Workplace Health System</u>

The Workplace Health System is introduced to Canadian companies by provincial ministries and agencies, Health and Welfare Canada's partners in the goal of integrating health considerations into the normal business routine. There are four components to WHS: the Small Business Health Model, the Corporate Health Model, Corporate Challenge and Evalu-Life. The first two offer a step-by-step method of introducing comprehensive health promotion programming into the workplace. These are based on a comprehensive needs assessment of all of the workforce, the setting up of a Corporate Health Plan and the implementation of a potentially wide range of programmes which target the three areas of influence (see below). The other two components can be used either as part of an overall health plan or on their own.

WHS has a number of advantages, namely:

* a practical, flexible formula for a health programme that meets employees' real needs;

* the benefit of experience: each component of the system has been tested at various worksites across Canada; and

* a means of coordinating the efforts of local health services groups, and for small businesses to share program costs.

The system was established with standard health principles in mind, so companies who use WHS should make sure that they:

* meet the needs of all employees regardless of their current level of health;

* recognize the needs, preferences and attitudes of different groups of participants;

* recognize that an individual's lifestyle is made up on an independent set of health habits;

Box 3. The Workplace Health System - Health and Welfare Canada

* adapt to the special features of each workplace environment; and

* support the development of a strong overall health policy in the workplace.

The health principles are linked to another important premise: that creating a healthy working environment requires addressing three broad "avenues of influence" which affect the way people feel:

* environment, including air, noise and light conditions, the quality of equipment, the type of work and the responsibilities, and relations with co-workers and supervisors;

* personal resources, including how much control employees have over their health and their work, the support they receive and the degree to which they actively participate in improving their own health;

* health practices, including exercise, smoking, drinking, sleeping and eating habits, and the use of medication or other drugs.

In short, to have a greater impact employers have to go further than simply providing fitness classes at noon, by offering literacy or other forms of skills-upgrading courses, workshops on quitting smoking or losing weight, stress management courses, financial planning seminars, counselling services and retraining.

Various resources have been developed and tested to support the implementation of both the Small Business Health Model and the Corporate Health Model. Though various aspects of the programme are still under development, the programme has been successfully applied in a wide range of small and large organisations, and has recently been applied in a developmental way to family farms.

(Adapted from Dooner, 1990/91)

6.6 Overcoming barriers to action on workplace health

In this section the strategies outlined above are integrated into practical action points for overcoming the barriers to actions on workplace health. The principal barriers identified in the survey are addressed - lack of facilities, lack of finance and lack of commitment.

6.6.1 Dealing with lack of facilities

The information supplied in the research programme tends to be quite sketchy on this point. However there are indications that lack of suitable facilities, be they in terms of buildings or equipment, can be overcome at least to some extent other than by means of the obvious solution of actually providing these facilities (which may be a costly route).

The solutions which appear to have been successful largely involve the use of facilities other than those owned by the organisation concerned with the health activity. A good example of this is to be found in the Fruit of the Loom-McCarters case study in Ireland (see Chapter 2), where a relative lack of on-site facilities prompted the company to use external facilities. In particular, collaboration with local health authorities in the setting up of health promotion weeks using some on-site facilities, but also those available to the local authority, were seen to be a successful approach.

It is also possible that citing a lack of on-site facilities should not be a serious impediment to health actions taking place in the first instance. It is likely that this problem being cited is more an indication of a lack of commitment, be it in terms of conceptual or resource commitment by the company concerned.

6.6.2 Overcoming lack of finance

It is important to note that many of the health actions outlined in the earlier Chapters do not necessarily involve large commitments of funding. Indeed some of them may involve no extra funding at all (particularly some of the organisational interventions which tend to take place in any event within companies, and which need a health component running through them, rather than the provision of extra funding). In addition, at least some companies profess that there are savings to be made in terms of reduced accidents, increased productivity and reduced absenteeism.

However, it is also clear that the provision of budgets overcomes many of the problems associated with establishing health activities. It is notable that from the regression analyses that the provision of health budgets was strongly associated with higher levels of health activities in the workplace. In addition, the provision of budgets (even if they are small) has also been cited as a means of overcoming a lack of financial resources for undertaking health promotion. Finally, it should be noted that the benefits available in both tangible and intangible terms would appear to be sufficient in many cases to over come the costs associated with establishing health actions in the first instance.

6.6.3 Dealing with lack of commitment

The issue of boosting management commitment to health actions in the workplace would seem to have been resolved among many of the respondent companies. Two issues are of importance here, the first concerns the generation or the creation of an adequate and comprehensive plan for the health action. Many respondents cited a lack of planning as a problem to management. The second concerns generating an increased awareness of the benefits to be gained from undertaking health actions in the workplace.

6.6.4 General comments on establishing health actions

Underlying much of the low levels of diffusion of health promotion activities in the workplace is the lack of suitable models or methods to enable them to become widespread. Well developed models for establishing workplace health actions would overcome many of the problems of commitment be it on the part of management or employees. The benefits to be gained from using well developed methods which involve the ultimate users of health promotion programmes at least to some degree overcome problems of lack of expertise within the organisation undertaking the activity.

Models such as the Health and Welfare Canada model offer particular promise in this regard. The HWC model emphasises the control of the company over the direction of the programme and stresses that the implementation of the programme depends upon the availability of sufficient resources. It also emphasises that structures which are set up to manage health actions within the workplace are designed on the one hand to increase organisational control (by both management and employees) and on the other to increase the skill levels necessary to implement the action. Through the support of external facilitators the key actors within the workplace acquire the skills and attitudes appropriate for the management of the programme.

The issue of lack of expertise also needs to be addressed. In many cases companies already have occupational health departments staffed by either physicians or nurses or other professionals. Yet expertise of these people appears to be lacking in some regards. This perceived lack of expertise is to some degree symptomatic of prevailing health culture which is treatment oriented rather than prevention oriented or health promotion in its orientation. Generally, this combination of traditional approaches and perceived lack of expertise underlines the need for appropriate training to be provided not alone to existing occupational health professionals but also to employees more generally.

6.7 So how are workplace health actions established ?

It appears that the best way to establish Workplace Health Actions involves planning, involves knowing what the problem is and involves commitment by management and workers. The examples in Box 1 illustrate the virtues of planning undertaken by a company as a whole. The problems which arose in the multinational as illustrated in Box 2 on the other hand illustrate the failure of apparently sound planning processes through a lack of commitment and a minimalist interpretation of what policies are. The activities outlined under the Health and Welfare Canadian model in Box 3 on the other hand illustrate how a comprehensive model which involves planning and problem identification can be used to promote health in the workplace.

The issue of participation and involvement is somewhat clouded by the results emanating form the survey. On the other hand many of the case studies outlined in Chapter two show that high levels of involvement occur in workplace where there is a high level of activity, even though the form of participation may be somewhat informal. On the other hand it is clear that the issue of commitment (as opposed to involvement per se) is very important. Many of the problems cited by respondent organisations related to lack of commitment be it on the behalf of management or on the behalf of the workforce. A graphic illustration of this is to be found in Box 2.

However the typical problems (resources, finance, commitment) faced by Workplace Health Action can be overcome. The examples provided in this chapter illustrate that with commitment and imagination, most, if not all of these problems can be overcome.

7. What Does It All Mean?

This chapter examines some of the most important themes to emerge from the research, particularly dealing with the issues of national differences, levels of participation, and levels of innovative workplace health actions. Also of interest is the issue of integration of health actions into a normal company activities and the future prospects for health actions.

7.1 Introduction

The information which has been gathered to date in this research programme is complex and wide ranging. The national overviews which were provided from the eight countries and from the survey and multinational case study results from seven of these eight countries (with the exception of Portugal) provide a rare insight into the conduct of health activity in the European workplace. There are of course major disparities between the countries which have taken part in the research. To some degree there are fault lines to be seen both on North - South dimensions and along economic development dimensions. However there is also a lot of common ground between the countries.

7.2 Legislation

In relation to the legislative background, it is clear that there are wide disparities between the countries in terms of current statutes. However, the Framework directives and its related directives in the area of occupational safety and health will promote a convergence of legislation over the coming years in EC member states.

In general, legislation does not specifically refer to workplace health promotion in any of the countries, though some of the clauses within legislation, although they are usually quite general in nature, do refer to issues which can be interpreted as involving health promotion. Perhaps the most obvious example of this, which exists in many national legislations, are clauses covering conditions of work and work organisation not being injurious to health. (The Irish Health, Safety and Welfare at Work Act, 1989 and the Dutch Working Conditions Act of 1980 are good examples).

Despite the convergence of legislation which will take place, it is unlikely that new legislation will refer specifically to health promotion. Nonetheless some of the newer concepts and some of the empowering aspects of legislation (such as rights of access to information for workers, rights of participation) will serve to create a more positive atmosphere in which health promotion in the workplace can develop.

7.3 The views of the major actors

The orientations of the major actors are generally positive towards health promotion in the workplace. To some degree this reflects many aspects of interactions between employers, trade unions and governments where laudable goals are proposed. However, the aspirations of the statements made by these parties tends to be somewhat in advance of the practice of policy on the ground.

While it is notable that there are few theoretical problems for most of the key actors in reporting positive orientations towards workplace health promotion, some concerns were expressed regarding the implementation of health promotion. Trade unions expressed concern at its potential use as a means of worker control (for example in the Netherlands and Ireland), while the perception that workplace health promotion could become a responsibility of employers as opposed to an adjunct to their more general human resource management policies was of concern in the case of the employers.

Governments do not seem to have taken on board the issue of health promotion in any serious way, either in terms of their legislation, or more particularly in the embodiment of legislation in terms of key agencies. Even within the existing systems there is often a myriad of agencies competing with one another to regulate workplace health action (the most complex system appears to be that of Germany, where the regulation of workplace in the former West Germany involved a multiplicity of agencies. The situation has become even more complex with the integration of former GDR agencies).

7.4 Lessons from the case studies

The concept of health promotion is a difficult one to grasp, partly because of its novelty and because it should involve multi-disciplinary, multi-method approaches to often multiple problems. It was notable that many of the multinationals viewed themselves as being involved in health promotion (to the extent that 'health promotion' departments existed, yet they did not appear to embrace a clear concept of health promotion - to the extent that these departments did not have formal linkages with occupational safety and health (see Hauss 1992 for more details). In addition, the difficulty of focusing on the promotion of good health as opposed to the clarity and ease of identification of ill health means that the concept has had difficulty in achieving widespread acceptance.

However, the case studies which were carried out in the first phase of the research programme (see Chapter 2), provide a wide ranging set of examples of how health promotion might be conducted within the workplace. These range from what might be termed relatively narrow but thorough health promotion programmes (e.g. Heartbeat Wales) to health promotion programmes which proceeded on a large scale and involved high levels of controversy (see for example the Italian case studies from phase one of the research). In addition, they provide examples of what can be done even in companies or areas not normally associated with extensive health policies, e.g. the Portuguese case studies (see Chapter 2 and Graca and Faria, 1991) and the Fruit of the Loom case study (see Chapter 2 and Wynne 1990).

Taken together, they provide a unique insight into the possibilities for health promotion in the European workplace. Furthermore, the case studies from both phases of the research constitute examples of best practice. They can be contrasted for example with many more limited examples of what has been termed health promotion, e.g. Wynne, 1989, Sloan et al 1987, Washington Business Group, 1987.

It is of course recognised that they do not constitute typical examples of health promotion in the workplace. It is probably impossible to state what a typical example might be since the survey findings indicate that true multi-disciplinary, multi-method health promotion programmes which focus on the promotion of good health are relatively rare on the ground.

However, some of the multinational case studies provide examples of well planned, coordinated and implemented health promotion activities in their workplaces. However, it would appear to be two types of approach to workplace health activities within multinationals. On the one hand, there are those who have a strong corporate health policy which is applied in all plants in all countries regardless of the provisions of various national legislations, and on the other there are those which have relatively weak corporate health policies and which apply examples of health promotion only in some of their plants.

The evidence from these latter types of companies would seem to indicate that this is not a policy which has been made explicit, rather it has happened that within some plants the multinational has adopted good practices and have not seen fit to export them to their daughter plants in other countries or indeed even within the same country. Perhaps one of the lessons to be learned from this pattern of health activity is that companies which are genuine multinationals (with vertical predict integration) as opposed to conglomerates operating in different countries, would seem to have stronger corporate health care policies and indeed stronger corporate policies generally.

The multinational case studies also provide insight into the mechanisms of participation in health promotion programmes. These are dealt with further in Section 7.8.

7.5 Lessons from the survey

The findings from the survey in seven member states provide a number of perhaps surprising findings. Perhaps the most interesting of these is that many actions which could have a health component in them are undertaken for reasons other than the improvement of worker health. This finding is particularly surprising in relation to actions which have a strong visible health component such as health screening activities. However it also indicates that in relation to activities which perhaps do not have such an obvious health component, such as shift schedule design or the design of working time generally, that there exists many opportunities for health to be inserted on to the agenda of companies.

Insertion of health considerations into the normal company processes concerned with planning, budgeting, controlling, manufacturing etc. should in many senses be the goal of workplace health promotion. It follows from the WHO guidelines on health promotion in the workplace and mirrors one of the aims of the Ottawa Charter, i.e. the creation of healthy public policy. The survey unfortunately provided relatively little evidence that such an integration of health into normal corporate activity has taken place.

Firstly, activities which might constitute normal corporate activity (such as the organisational interventions) have tended to take place at a relatively low frequency and of those many took place without a notable health component. Indeed these were among the least prevalent health activities investigated in the survey in terms of the level of health concerns.

Secondly, there was evidence from examining the role of occupational health departments that they tend to operate on the margins of company life. For example, one of the most striking findings in the entire survey concerned the fact that the presence of occupational health department only predicted levels of involvement in health activities of occupational health staff. In other words they did not predict the involvement of management, workers generally health and safety representatives or trade union representatives in health activities. The extent to which this is a comment on the limitations of occupational health personnel or a comment on the resistance of companies to the integration of health concerns into their normal operations is a matter for debate. However, there is evidence to suggest that the concerns of occupational physicians in particular tend to be very narrowly focussed.

To some degree this reluctance to become involved reflects the wider rivalries which exist between public health and occupational health. One view of workplace health promotion is that it constitutes the export of public health practices into the workplace. Against this background it is important that potential tensions between occupational and public health are recognised, and that an effective strategy for integrating their respective concerns be devised.

Whereas there were quite large national differences in relation to some of the indices of health actions and forms of participation, perhaps the most interesting finding concerned the role which companies health characteristics had in predicting levels of health activity. This class of variables was possibly the strongest predictor of health activity within respondent organisations. In short, the existence of health policies, health budgets, occupational health departments and health and safety committees had a major role to play in the creation and establishment of workplace health activities.

Backing up findings with regard to health activities which had a health component in them was the information concerning prompting factors for health actions in the workplace. Whereas it might be supposed that health actions take place largely for the purpose of improving health, the findings from the survey indicate that this is not so. While health improvement was certainly a factor, it was not the only factor, nor even the largest.

In fact the findings showed that issues such as legislation, the improvement of company morale and the existence of personnel problems also had major roles to play in the establishment of health actions. These findings have considerable significance for what might be termed the marketing of health promotion in the workplace. They indicate that it is not necessary to justify health actions solely on the basis of health improvement. The range of benefits which emerged from engaging in health actions tend to back up this point, where health improvement was certainly seen as one of the benefits to be gained. Many other benefits were also outlined. These included improvements in productivity, improvements in staff morale and perhaps most notably, improvements in industrial relations.

7.6 The Situation of SME's

It was abundantly clear from the survey that one of the most powerful predictors of health activities in the workplace was the size of the organisation. This findings applied to all seven countries surveyed. Large organisations undertake more health activity than SME's. This may be due to greater levels of resources, be they financial, human or time.

It is also clear that there are few models available which enable SME's to effectively undertake health actions in the workplace. There are some models of occupational health services applied within the EC which may be appropriate for consideration which involve group practices or groups of SME's coming together to jointly fund an occupational health service. However, another model which is of great potential is the Health and Welfare Canada model which is specifically targeted one of its Workplace Health System Programmes at SME's - the Small Business Programme. In this model an agency external to the SME liaises with a group of SME's, (perhaps operating within a particular community) and develops a health promotion programme based on the needs analysis of the participating group of SME's. In this model it is possible for individual SME's to opt in or opt out to the extent that they feel able, based on the resources and their commitment.

There is a commitment at EC level to encourage the growth and activities of SME's in the economic sphere. It is also appropriate that there should be a commitment at EC level, given that the problems that SME's face are common across countries. To promote health activity within SME's, new models of uptake of health promotion need to be considered. The EC could do worse than examine models such as the Health and Welfare Canada model as potential vehicles for increasing health promotion within SME's.

7.7 National differences

Not unexpectedly, the survey provided evidence of quite large scale national differences in levels of health activity. The factors which contribute to these differences are to some degree obvious, and they include levels of economic development, the differing political histories of the states involved in the research, differing levels of experience of workplace health actions, the size of enterprises, the sectors in which they operate, differing industrial relations traditions and more broadly based cultural differences.

There may be an important lesson to be learnt from the differing legislative backgrounds against which workplace health actions take place. In Germany, for example, many of the provisions of law are quite specific (even if not demanding), and the existence of well defined structures and procedures within workplaces provides a supportive backdrop for the establishment of workplace health action. This situation is reflected in relatively high levels of workplace health action.

By contrast, in the UK and Ireland legislative provisions (at least in the past) tend to make the establishment of workplace health actions of any description voluntaristic. In both of these countries workplace structures for dealing with any workplace issue, not alone those concerning health are far less developed. It is perhaps unsurprising that both the UK and Ireland reported some of the lowest levels of health activity.

There is a danger however in believing that 'compulsory' health actions (such as those in Spain) which lead to higher levels of specific activities are necessarily indicative of benefit. Interpretation of the effects of these high levels of activity would suggest that the health screening activities in Spain, for example, have limited effects on the health of workers. In addition, such compulsory activities can have the effect of shifting scarce personnel and resources away from more fruitful activities.

Whatever the causes of national differences in workplace health activities, the responsibility for the remedies to these differences clearly lie at the level of the EC, national governments and to a lesser extent, companies and organisations operating in more than one country. The EC can play a role in ensuring a level playing field in terms of legislation and in terms of providing support networks and vehicles for information and 'technology' transfer.

National governments also have responsibility for 'levelling up' the prevalence of workplace health actions. In doing so it is recognised that governments of less well developed economies are in something of an ambivalent situation - whereas they must seek to encourage workplace health activity through the Framework and related Directives and the implementation of the Social Chapter, they must balance these commitments against the possibility of penalising native companies with additional costs which make them less competitive.

Finally, companies who are active in the field of workplace health and who operate in more than one country can also play a role. The evidence from the multinational case studies suggests that much more could be done to ensure the 'export' of good practices in this arena from one country to another.

7.8 Lessons on participation

The findings regarding levels of involvement of the six major workplace actors were also striking. Management were reported to have by far the highest levels of participation, both throughout all stages of the life cycle of workplace health actions and also in terms of the level of involvement which characterised their participation.

It would seem from these findings that there is some way to go in terms of involving workers in the planning and implementation of workplace health promotion activities. In addition there were relatively low levels of involvement of health and safety representatives, occupational health staff and trade unions begs the question of their level of commitment to health actions in the first instance. Again these findings may reflect resistance on the part of management to involve these groups, but the extent to which they wish to become involved must also be a matter for debate.

Indications from the case studies suggest that participation practices vary far more between countries than between multinationals (see Hauss 1992).

In countries where worker participation is relatively institutionalised, e.g. Germany, similar levels of participation takes place in different countries. Equally, where participation is less institutionalised e.g. Ireland and the UK similar levels of participation also take place. The relative effectiveness of these different forms of participation in relation to health activity is not clear - while it may appear that the more institutionalised forms are preferable this was not necessarily borne out in the case studies. Institutionalised forms of participation may become stereotyped, while more informal forms such as that described in Box 1 in Chapter 6, appear to be able to function well (see also Wynne and Clarkin 1991).

7.9 Future prospects for workplace health promotion

Any assessment of the future prospects for workplace health promotion must take cognisance of the positive factors which tend to support health promotion, the neutral factors and the negative factors which militate against the development of workplace health promotion. Knowledge of these factors is to some degree already available, but this research programme has gone some way towards accurately specifying the nature of these factors.

On the positive side, it is clear from this research programme that there are many examples of good practice in the area of workplace health promotion. If nothing else, this research demonstrates the possibilities for health promotion in the workplace in a vivid way. Also on the positive side, the survey evidence presented in this report amply demonstrates that respondents believe that there are benefits to be gained from engaging in workplace health activity. These benefits occur not only to the individual, in terms of improved health and morale but also to the company in terms of hard measures such as reduced absenteeism and improved productivity, but also in terms of softer measures such as improved image, and improved staff morale.

The principle neutral factor in assessing the prospects for workplace health promotion concerns legislation. Whereas much of the existing legislation is only at best neutral in that it does not prevent the conduct of workplace health promotion, it would appear that many of the legislative changes now under way will offer a firmer background against which workplace health promotion can take place. It should be noted though, that as far as this author is aware, none of the legislative changes proposed actively support workplace health promotion.

Against these positive and neutral factors, a number of problems persist which hinder the development of workplace health promotion. These include a lack of awareness amongst most of the key actors. Awareness in this regard refers to issues as levels of awareness of health promotion - as a concept, in terms of its possibilities, its benefits and also in terms of what it entails. It is hoped that the current research will go some way towards addressing this awareness gap.

Another problem is concerned with the skills, particularly of the professionals but also of the company personnel who would be involved in workplace health promotion activities. The professions themselves would appear to cling to a narrow definition of their role which is largely focused on health screening activities, safety activities and to some degree prevention activities. On the basis of the evidence from the research programme, it would appear that they have neither the awareness, the skills or perhaps the motivation necessary to establish workplace health promotion.

Further problems concern the marketing of health and health promotion in the workplace context. In this regard the current research offers what may be a key marketing tool, i.e. the realisation that workplace health promotion programmes do not need to be marketed solely on the basis of health, but can also be marketed in terms of benefits to be gained from engaging in such activities. The role of some of the factors which may prompt health action in the first place are also of importance to the marketing process - these include legislation, morale problems, health problems, and also a fairly wide range of more commonplace workplace problems, upon which workplace health action seems to have a beneficial effect (industrial relations problems are most notable in this regard).

Most countries cited a lack of finance as being a problem. However, as pointed out in the previous Chapter, this may be more apparent than real, since many health promotion activities cost relatively little, and the 'harder' benefits to be gained may justify the cost of health actions in any event. Allied to the lack of finance is a lack of facilities to conduct workplace health actions. This is also a problem which can be overcome with a little creativity. Use of off-site facilities is often used as a means of overcoming this problem. In addition, many health promotion activities do not need physical facilities to take place, for example organisational interventions aimed at the psychosocial work environment.

The attitudes of the major actors in workplace health can be viewed as having both positive and negative facets. Whereas most actors profess a positive orientation towards workplace health there is a considerable gap between these attitudes and practice. They support the concept of health promotion in theory, but have shown little inclination to commit sufficient resources to the establishment of workplace health promotion.

Another problem concerns the roles of the various institutions from the EC Commission and governments down to the levels of employers, trade unions, the professionals and the agencies which enforce occupational health and safety legislation. In many countries such as Germany, the UK, Ireland and Italy the roles on the ground of these various actors seem confused. In addition, there seems to be considerable under resourcing of these agencies (particularly the enforcing agencies), which leads to considerable inertia in the system if workplace health promotion is to be encouraged.

Despite this fairly long list of problems facing the establishment of workplace health promotion, it would appear that future prospects are improving. For example, the current research provides a wealth of raw material to help in the process of increasing awareness. It provides information which is of use in marketing of health promotion in the workplace and has the benefit of defining many of the problems which face workplace health promotion and the major actors who may be involved in it. Furthermore, health promotion in its widest form is coming on to the agenda of many of the international agencies (WHO, ILO, EC), and this is reflected in increased activity in the area both inside and outside of the workplace (e.g. the Healthy Cities Programme, European Year for Occupational Health, Safety and Hygiene at Work). On this basis it could be said that future prospects are improving.

On the other hand major problems still remain which are not going to be solved by publishing reports such as this, or indeed by already existing international programmes. These concern the commitment of resources to such issues as improving skills of the personnel concerned with workplace health promotion, developing awareness among both professionals in the workplace and the workforce at large about health promotion, and the role of institutions (e.g. health and safety authorities, health services, professional groups) and the social partners in the whole process. Until these problems are seriously addressed, (and they are subject to recommendations in the final chapter), it is likely that workplace health promotion will only move ahead at a slow pace among companies who are already initiated into the benefits to be gained from engaging in such activities, while the vast majority of workplaces and workers will remain completely unaffected by health promotion at work.

8. What needs to happen ?

This short chapter proposes a number of conclusions and recommendations from the research programme to date. These involve such issues as commitment, training, information dissemination, enforcement of legislation and transferring health issues into normal company operations. The recommendations which will be made will be of three types: actions which could be taken at EC level, actions which are most appropriately undertaken by national level agencies and actions which can be taken by management and unions at company level.

8.1 Introduction

In making recommendations from the wide body of research conducted under the European Foundation Programme three issues should be borne in mind. The first concerns the comprehensive nature of the research - to date this programme has provided the most comprehensive transnational body of information available in the EC. The recommendations therefore, while acknowledging the limitations of the research, carry considerable weight.

Furthermore, the fact that the research provides a first glimpse of the state of the art of European workplace health promotion means that the recommendations have particular significance for policy and practice at EC and national levels.

Finally, the fact that the research combines both survey and wide-ranging and rich case study material enables the recommendations to be more than suggestive. In particular, recommendations aimed at the company level carry considerable weight. This is because the insights to be gained from case studies of good practice in conjunction with the fact that the organisations who completed questionnaires for the survey were likely to be those with most experience of workplace health action provide convincing evidence of the benefits to be gained from workplace health promotion.

The recommendations have been drawn from materials from the entire project and not just from the current phase of the research. In essence this means that material from the academic literature, the attitudes of the major actors, legislation in the eight EC countries, case studies of good practice, the survey and the recommendations made by national researchers have been combined to generate the recommendations.

It is beyond the scope of this report to propose recommendations for each of the eight countries which took part in the research, as this is a task best left to each of the national researchers. (In fact this has already been done in most of the national reports - see Aravidou et al, 1991; Clarkson et al, 1991; Garzi, 1991; Graca and Faria, 1991; Grundemann and Ellis, 1991; Hauss, 1991; Moncada, 1991; and Wynne and Clarkin, 1991). What has been done, however, is to point to where it is appropriate for the EC to become active in the area of reducing national differences in activity.

The recommendations have been defined at three levels: the EC and governments, institutional agencies (this refers to employers organisations, trade unions, government agencies and professional bodies) and individual organisations or companies. It should be noted that the recommendations are not given in order of priority. In the recommendations emphasis has been placed on identifying the agencies who should implement them and the process whereby they can be achieved. A summary of these is provided in Table 8.1 below.

In essence, the major stakeholders in the workplace which are found in the eight nations have been addressed by these recommendations. A further stakeholder exists in some of these countries - the insurance companies. Recommendations have not been targeted at insurance companies on the grounds that they do not operate in all of the EC countries. However, the recommendations which concern the institutional stakeholders are largely applicable to them.

Table 8.1 Summary of the recommendations

Issue	Agencies affected
Dissemination	EC, Trade Unions, Health and safety bodies, Employers organisations, Academic community
Research	EC, Governments, Trade Unions, Health and safety bodies, Employers organisations, Academic community
Training	EC, Governments, Trade Unions, Health and safety bodies, Employers organisations
Legislation	EC, Governments
Enforcement	Governments, Health and safety bodies, Employers organisations, Trade Unions
Methodology	EC, Governments, Health and safety bodies, Employers organisations, Trade Unions
SME's	EC, Governments, Health and safety bodies, Employers organisations, Trade Unions
Health structures in the workplace	EC, Governments, Employers organisations, Trade Unions
Participation	EC, Governments, Employers organisations, Trade Unions, Academic community

Underlying all of these recommendations is the belief that workplace health promotion is beneficial for all of the parties concerned. This belief is disputed by many sources, some on the grounds of their health improvement efficacy, and others on the grounds of their economic cost effectiveness. It is beyond the scope of this report to deal thoroughly with these issues, but it should be noted that whatever the theoretical implications of effectiveness, it is beyond dispute that many companies have seen fit to introduce workplace health promotion. Furthermore, these companies report benefits in precisely the areas of health improvement and of cost effectiveness. It is against this background that the recommendations have been framed.

Recommendation 1 - Information dissemination

Information regarding health promotion at the workplace needs to be collected and actively disseminated at EC, national and company levels.

Background: There is a wide body of evidence collected in the current research programme and in other studies which emphasise the need for information dissemination. In particular, there is a widespread lack of awareness amongst almost all of the workplace actors about potential health actions which they may take in the workplace. To some degree it is true that there is not a lack of sufficient instruments and methods for undertaking workplace health promotion (this is being noted particularly in the German report for example - Hauss 1991). But there is an obvious dissemination gap in so far as many major actors in the workplace continue to use restricted models and concepts of what health actions can mean.

What the EC should do: It is clear that the EC can have a major role to play in the dissemination of information about models of best practice for workplace health action. To some degree, this has already been addressed with 1992 being designated European Year for health and safety and hygiene in the workplace. However, many of the dissemination activities which are being undertaken in this year, at least as far as this author is aware, appear to have no lasting presence beyond 1992. It is recommended that the European Community set up an information distribution network (on paper and/or IT based) to promote awareness about the possibilities about workplace health action.

It is also of interest that the World Health Organisation has designated as one of its collaborating units, a centre for the dissemination of case studies of good practice in workplace health promotion (BKK 1991). However, at this point the only case studies contained within the database provided by BKK concern German examples. There is a need for information gatekeepers such as this to be encouraged, to have their services extended and to be properly funded.

What the institutions should do: In relation to trade unions, health and safety bodies and employers organisations, there is a dual role in relation to dissemination. As dissemination is primarily about increasing awareness, these bodies must be seen as being both vehicles and targets for dissemination efforts. In particular there is ample evidence from the survey and also from earlier work within the current research programme that trade unions, health and safety bodies and employers, tend to view workplace health within the narrow confines of traditional health and safety practice. Furthermore, there is also evidence that there is a gap in their awareness of what is actually possible in relation to workplace health promotion. In this regard, these bodies, should become the targets of dissemination activities.

It is also true that the trade unions and health and safety bodies and employers should become important vehicles for dissemination of information on workplace health promotion. To some extent they are already act in this way in relation to narrower health and safety issues. In principle once they have sufficient information at their disposal, they should also act as gate keepers and disseminators for information on workplace health promotion.

What the academic community should do: The academic community to date has shown limited interest in workplace health promotion, perhaps because in many countries in the EC (for example, Greece, Portugal, Ireland, Spain) there is little enough evidence of genuine workplace health promotion actually taking place. However, academics in the larger countries do have some experience of investigations into workplace health promotion.

It would appear that the dissemination of the research work which is being carried out within these countries is not as widespread as it could be. This is evidenced in the lack of awareness among some of the major workplace actors regarding workplace health promotion. It is recommended that the academic community contribute in a full way to the creation and maintenance of dissemination networks. In addition, information which they hold should be made more openly available to the key workplace actors.

Who should take the lead ? It is clear that information regarding health activity in the workplace resides largely at the EC, government and institutional levels, and that a gap exists at the level of the enterprise. It is therefore appropriate that the EC, government and institutions should take the lead in information provision initiatives.

Recommendation 2 - Research

Research into the prevalence of health actions, their method of establishment, the costs and benefits of health actions, the factors which promote workplace health action and the methods of overcoming barriers needs to be undertaken.

Background: As noted previously there is a chronic lack of research in many countries in to the extent of health actions in the workplace and into the factors which facilitate or constrain workplace health promotion. In addition, issues such as the lack of practically applicable cost benefit assessment methodologies, the general lack of information about factors which promote the establishment of health promotion in the workplace, and practical means of overcoming barriers to health promotion in the workplace have not been the subject of sufficient research.

More generally, the lack of evaluation of currently existing workplace health promotion programmes is also an area to which the research community could profitably contribute to the promotion of workplace health. A further issue in which the research community would have a major role, concerns methodologies for the introductions of workplace health promotion programmes. To some degree work has already been done in this area, for example, the WEBA Methodology has been under development in the Netherlands and also the worksite health promotion programmes of Health and Welfare Canada which relate to large and small companies and lately to the family farm sector (Dooner 1990/1991) offer promising methods from which flexible and adaptable approaches to workplace health promotion in Europe could be developed.

What the EC and governments should do: It is plain that research needs to be undertaken at the level of the European Community as many of the individual states do not have sufficient research funding to undertake major research in this area. It would appear to be appropriate that the European Community could at least part fund such research, perhaps under the auspices of its workplace health and safety activities or through its public health activities. Where possible, governments should also fund such research.

What the institutions should do: National Health and Safety Authorities could also play a role in providing funds for such research or indeed undertake such research themselves. In view of the legislative trend towards the empowerment of workplace actors to undertake health and safety activities, it makes sound sense for such bodies to provide them with the tools for adequately carrying out the health and safety function. A further source of research funding should be the bodies who carry the insurance risk for workplace health and safety claims, be they private or public.

If the results of such research are to be applied in practice it is important that the employer organisations and trade unions are involved in the research process itself, be it through providing funding or through having a role in steering committees for such research. The participation of the social partners would ensure that the research results were disseminated and applied in practice.

What the academic community should do: The academic community has a major role to play in the development of research along the lines suggested. It is recommended that they should do so in partnership with the social partners rather than as an academic exercise in its own right, if the results of research are to be applied in practice.

Who should take the lead ? Research is largely a centrally organised and funded activity. Against this background the EC, government and institutions should take the lead in developing, funding and implementing appropriate research.

Recommendation 3 - Training

Training programmes concerning workplace health promotion for practitioners and professionals need to be instituted. Access to currently available training programmes needs to be increased.

Background: There would appear to be considerable skill shortages in relation to workplace health promotion (if not also in relation to more general occupational health and safety) in most of the EC member states which took part in this current research. It is clear also that the current round of updating of occupational health legislation taking place as a result of the completion of the internal market will provide challenges to the training establishments within each of the member states. Even without these legislative changes, there is considerable evidence that there is a shortage of sufficiently trained personnel and a shortage of training infrastructure in many countries to undertake even the task of traditional occupational safety and health.

What the EC and governments should do: Given this skill and training infrastructural shortage and the background of EC inspired legislative changes, it would appear appropriate that the EC should have a major role in stimulating training within each of the member states. Although it is not feasible to expect the EC to provide training itself, it is appropriate to expect them having inspired changes in legislation in the first instance, to provide sufficient backup to enable training programmes to move forward, to meet the demands of new legislation.

This could be done for example through the development of training modules at an EC level which could then be supplied to each of the member states for application within their own training infrastructures. These training modules would be based on EC directives and would also extend to cover more explicitly the issues involved in workplace health promotion.

The EC could also address the problem of lack of training infrastructure in its own right. It appears certain that in some countries at least, that regardless of any changes in legislation which may take place, they already have difficulties in meeting the needs for training in this area. It also appears to be the case that funding to sufficiently increase the resources of training infrastructure within many of the member states will not be forthcoming on the scale which is needed to address the magnitude of the problem.

In this context, alternative approaches to training are called for. The possibilities of distance learning using Information Technology and Telecommunications are already well advanced in many member states but the possibilities of these technologies for use in the area of health and safety training or indeed health promotion training for the workplace have not yet been explored. The EC already support the development of technological infrastructures for such training through programmes such as DELTA and RACE, but crucially, the EC appears to have no role in relation to the actual development of courseware. It is recommended that such a role be developed, in order to support the member states in providing both training infrastructure and training courseware.

Within each of the member states governments obviously have a role to support the development of training within their national jurisdictions. It is strongly recommended that they take a fresh look at current training infrastructure and courses, to update them to include on the one hand, new materials, targeted at workplace health promotion, and on the other to increase support for infrastructure which can have the effect of greatly increasing the dissemination of technological skills and concepts in the area.

What the institutions should do: It is in the interest of both trade unions and employers to oversee the further development of health and safety training in the workplace. It is recommended that their own actions in this area be supportive of EC and government actions and also that they actively support the incorporation of wider health promotion issues into current training practices.

Who should take the lead? Though training programmes are implemented best at the level of the enterprise, the problems of infrastructure within many countries indicates that the EC and governments and their agents have a crucial role to play in the development of training. In particular, central agencies such as the EC should play a role in the development of new approaches to training.

Recommendation 4 - Legislation

The provisions of legislation need to be strengthened in order to provide a more supportive environment in which workplace health promotion can become established.

Background: During the course of 1992, there are wide ranging changes in legislation due to take place as a result of the completion of the internal market through the implementation of the Framework and related Directives. There would appear to be room for strengthening the provisions of legislation both at EC level and national level in relation to health promotion specifically as opposed to workplace health and safety in general.

What the EC and governments should do: In the review of the legislation presented by the national researchers, most strongly make the point that much of current legislation is not explicitly supportive of workplace health promotion, even though it does not prevent it from taking place. Given the findings that legislation can act as a major prompting factor for workplace health action it would be appropriate that more specific provisions concerning workplace health promotion are encased in EC and national legislation if workplace health promotion is to be actively encouraged.

It is unlikely that traditional style legislation would be appropriate, given the fact that health promotion is not solely a responsibility of the workplace. The possibility of providing incentives to companies by means of legislation may be a more promising approach.

Whether these changes in legislation should take place in the context of occupational health and safety law, in relation to public health law, or both is a matter for debate. In any event, it appears clear that the provisions of public health legislation and occupational health legislation should be more closely integrated.

Who should take the lead? The drafting and enactment of legislation appropriately takes place at the national and EC levels. The EC, as a stimulus for change in legislation and governments, as the sovereign powers have the responsibility to take the lead in this area.

Recommendation 5 - Enforcement of legislation

Currently existing legislation needs to be more rigourously enforced if workplace health action is to become more widespread.

Background: In most of the EC member states it is recognised that the levels of enforcement of current occupational health legislation are far from complete. In general, legislation tends to be adhered to and enforced in larger companies but there seems to be a considerable problem in establishing health actions (or indeed health and safety actions) within SME's.

The traditional approach to enforcement concentrates on the provision of an inspectorate which oversees companies covered under occupational health and safety legislation. However there has been to some degree a change of this approach whereby workplace actors have been empowered with the discretion to act as "surrogate" forces of legislation (see the Irish Health, Welfare and Safety at Work Act, 1989 for example).

It would also appear to be the case that governments are moving from legislative positions which involve coverage of only a limited portion of workplaces to legislative positions whereby most if not all workplaces are covered. Against this background it seems that the only feasible method of enforcement is to evolve responsibility for enforcement down to the level of the workplace itself.

What governments and institutions should do: It is therefore recommended that national governments take measures to empower key workplace health actors such as management, trade unions, health and safety professionals to enable them to have the skills necessary to comply with the provisions of law. (These steps should be reinforced by the practices of national health and safety agencies). It would appear that the best method of doing so is to ensure that the responsibility for occupational health and safety resides with management and that policing of health and safety legislation resides with workers representatives.

The enforcement approach of most countries has been characterised by penalising workplaces which do not conform to the provisions of law. This philosophy is entirely appropriate in relation to preventive and protective actions, but may not be feasible in relation to health promotion activities. (As outlined previously, there is a tension between the ways in which occupational health and safety are dealt with through legislation and the principles and responsibilities for health promotion). Though currently health promotion activities in the workplace are not explicitly required by law, it would appear that the best means of ensuring that legislation is used to promote workplace health promotion activities is through the encouragement of workplace actors to engage in such activities.

At minimum this should involve information campaigns to raise awareness of the possibilities of workplace health promotion and on the other hand to provide guidelines and clarification of currently existing legislation which reinforce the possibility of development and establishment of workplace health promotion in the workplace. In particular, the clauses in many national legislations which deal with conditions of work and the organisation of work, so as not to be injurious to health could be expanded to cover the kinds of health promotions activities which have been outlined here.

A further approach should also be considered which concerns the provision of incentives specifically for the development of workplace health promotion activities. These could take the form of tax incentives or perhaps worthwhile competitions for companies which engage in health promotion activities. This approach would seem to be warranted particularly if health promotion is not viewed as a compulsory activity.

Who should take the lead ? The failure to provide adequate resources for the enforcement of legislation is essentially a failure of political will and commitment. National governments carry the prime responsibility for this failure and should take the lead in ensuring adequate enforcement of legislation.

Recommendation 6 - Methodology for the establishment of workplace health promotion

A flexible, widely applicable and easy to use methodology for the establishment of workplace health promotion programmes needs to be generated.

Background: If the current research has identified one single issue of importance, it is that there is no widely agreed methodology for the establishment of workplace health actions. Though such methodologies do exist (at least to some degree), they have not received sufficient support from workplace key actors to enable their widespread application.

Currently available methodologies within the European Community include such examples as WEBA, and the health promotion programmes of many of the multinationals, for example 'Live for Life Programme' of Johnson and Johnson. In addition, there is also the major programme undertaken by Health and Welfare Canada in the provision of health promotion services within workplaces which would seem to be the most comprehensive of methodologies currently available. (This methodology provides for workplace health promotion in large companies, in small companies, and indeed in the agricultural sector and is explicitly concerned with health promotion rather than with health and safety only).

It is interesting that the mode of development of such methodologies has always taken place within context of large organisations and **institutions**. In developing the Canadian model, the Canadian Federal government have provided the funding and the impetus for the development of this methodology. In the Netherlands, the Dutch government have provided funding and encouragement to develop the WEBA methodology and even within the multinationals, they have had the resources to develop their own internal methodologies at headquarters.

What the EC and governments should do: If workplace health promotion is to be taken seriously in Europe, it is imperative that a convincing methodology for the establishment of workers health promotion in European workplaces be established. The logical source of funding and impetus for such a development should be at the EC and national levels. It is recommended that this take place and it is suggested that the Health and Welfare Canada methodology could be used as a basis for developing the European methodology, as this is a broader based model than any of the others available at this time.

What the institutions should do: It is important that all of the key actors are involved in the development of such a methodology - health and safety bodies, employers organisations, trade unions and academics. All of these actors should have a role to play in the development of such methodology, if it is to achieve widespread acceptance.

It is further recommended that in order to provide impetus for such a methodology, a series of demonstration or flagship projects be set up in each of the member states which would be used as vehicles for the development of methodology. Already, there exists a network of people who have sufficient skills and contacts to apply such a methodology and to run such demonstration projects, through such activities as the current research programme. These projects should function not only as vehicles for the development of a methodology, but should also be used for the purpose of raising awareness among the community about the possibilities and benefits of workplace health promotion.

Who should take the lead ? The development of an appropriate model for establishing workplace health action needs to be a co-operative process between all of the relevant stakeholders. However, given the difficulties of inserting health promotion onto the agenda of companies it is appropriate that the EC and governments set the pace in the development of such a model in terms of leading the initiative and the provision of (at least part) funding, but that the social partners should be involved in the actual development of the model.

Recommendation 7 - SME's and health promotion

Methods to encourage SME's to become involved in workplace health action need to be instituted.

Background: It is recognised in most circles that the diffusion of health and safety activities into small to medium size enterprises and particular into very small enterprises has been problematic in all member states. Even where legislation exists compelling certain kinds of health activity within small to medium enterprises, the results of this legislation have been disappointing (see for example Spain, where the mandatory use of health screening in companies of more than a hundred workers, has led to a minimalist interpretation of what health in the workplace should pertain to).

What the EC and governments should do: The wide ranging problems faced by SME's have long been the subject of concern at EC and governmental level. In relation to health activity, typical problems faced by SME's include lack of resources and lack of skills. It is recommended that EC and governments should provide a package of infrastructure and incentives to enable SME's become more involved in workplace health action.

Notwithstanding earlier comments on the availability of sufficient numbers of skilled personnel to enable workplace health promotion to become more widely diffused, there are models of approach to SME's which hold promise and which overcome some of these problems. In particular, programmes such as the Health and Welfare Canada Small Business Programme are possibilities in this regard. In addition, some of the models of group practice of occupational health can be found in some of the EC member states (eg. the Netherlands, Belgium) could be pursued more vigorously (see Rantanen, 1991 for details of these approaches to workplace health). The use of the facilities and skills for occupational health and safety of large companies by neighbouring small companies can also be further pursued. Finally, consideration should be given to providing incentives to SME's to become involved in workplace health action.

Who should take the lead ? National governments and the EC should take the lead in involving SME's in workplace health promotion, given their interest in encouraging the development of SME's.

Recommendation 8 - Health structures in the workplace

The establishment and encouragement of health and safety committees and occupational health departments, the provision of health budgets and the framing of health policies needs to be actively promoted within companies.

Background: Some of the most striking findings from the survey data concerned the power of company health characteristics such as health budgets, health policies, health and safety committees and the presence of occupational health department to predict levels of health activity. It would appear that the existence of such structures in the workplace provides much stronger possibilities for the occurrence health promotion activities. Accordingly, where they are not present it is strongly recommended that any organisation who wishes to undertake health promotion, at minimum begins with the defining a health policy, followed by the provision of a health budget and the designation of appropriate roles to carry out this policy.

What the EC and governments should do: At the EC level these findings have implications for the ways in which occupational health directives are framed. In particular, emphasis should be placed on the building up of appropriate structures in the workplace be they in terms of health and safety committees or in terms of occupational health services to support the development of health promotion in the workplace.

For governments a dual role could be envisaged. On the one hand when framing national legislation it is important to include the provision of such company health structures as part of the provisions of legislation. An interesting example, in this regard, is provided by recent Irish legislation which on the one hand (unfortunately) removed the obligation to provide health and safety committees but on the other obliged companies to produce 'safety statements' which outline the targets and activities in the health and safety area for all companies. (These 'safety statements' allow for the possibility of health promotion measures in response to perceived health problems in the workplace). Such a muddled approach to the establishment of these structures is likely to lead to a failure to adequately develop health promotion in the workplace.

What the enterprise should do: For companies and organisations the implications are clear. If they wish to seriously address the issue of health promotion they must actively define health policies, actively provide health budgets for them to be implemented and actively provide staff with the appropriate roles and skills to support the development of health promotion in the workplace. More generally for employers, organisations and trade unions it is recommended that they seek to encourage the development of such structures.

What professional organisations should do: For professional organisations in particular for those who cater for occupational physicians, a careful examination of their role is called for. Whereas the findings from the survey indicate quite clearly that the existence of occupational health departments is associated with some workplace health activities, it is also notable that occupational health departments tend to be associated only with the activities of occupational health personnel. In other words, it would appear that the predominant mode of action of occupational health personnel would appear to be that of an expert who "does" health to the workforce. Since this expert oriented traditional approach to health is inimical to the principals of health promotion a re-examination of the role and training of such professionals is called for. If health promotion is to be taken seriously by these bodies, it is essential that they move beyond traditional concerns of treatment to those of prevention and health promotion.

Who should take the lead ? The establishment of appropriate workplace health structures is essentially a problem to be found at enterprise level. While national legislation may dictate that some structures must be present in the workplace, the mere presence of these structures does not ensure meaningful health activity. Accordingly, the social partners carry the prime responsibility for ensuring that workplace health structures exist and also that they work effectively.

Recommendation 9 - Participation in workplace health actions

Methods of achieving effective participation by employees in workplace health actions need to be promoted by the major stakeholders in the workplace.

Background: The findings from the survey are somewhat equivocal with regard to participation. On the one hand it was clear that management were involved far more than other categories of actors in all stages of the development of health actions in the workplace. It was also apparent that management had by far the highest form of involvement when compared to the other actors. Perhaps because of the size of management involvement there was something of a "floor" effect with regard to the involvement of other key actor groupings. In any event, participation by workers and by occupational health staff tended not to be strongly associated with the occurrence of health actions.

On the other hand, theory would tend to say that higher levels of participation mean that more successful health activities are undertaken. There was some evidence for this from some of the multinational case studies where participation, if not always at the highest form, tended to characterise more successful activities within the multinationals. It is notable for example in the case of one of the Irish multinationals case studies (see Box 1 Chapter 7), that the involvement of workers is managed through an informal form of participation which could largely be characterised as consultation. Though this type of participation may not approach the highest form of participation, it nonetheless seems to be both culturally appropriate and successful in assisting the development of health actions in the workplace.

Recommendations are difficult to frame against this ambivalent background of the role of participation. However it is clear that some level of participation is appropriate in the development of workplace health actions, as evidence from both the multinational case studies and the other case studies indicates that worker involvement is of importance. In addition, evidence from the survey indicates that involvement of occupational health staff in particular is associated with higher levels of at least some activities.

What the EC and governments should do: At EC and national level, minimum requirements for participation of workers in the health actions which affect them are, and should continue to be safeguarded in legislative provisions. For employers organisations, and trade unions it is recommended that they implement already existing procedures but furthermore that they examine closely the results of current practices of employee involvement with a view to establishing the most effective form of involvement in relation to health actions in the workplace.

What trade unions should do: One of the most notable findings from the survey was that trade union influence on workplace health activity was minimal. It is strongly recommended that trade unions take a much more active role in supporting and initiating workplace health promotion activities.

What the academic community should do: For the academic community it would appear that there is a need for further research in this area. The received wisdom that high levels of involvement are better than lower levels may not necessarily be true. Further light needs to be shed on this issue and in particular on what might be termed informal forms of participation.

Who should take the lead ? Provisions for participation exist to a greater or lesser extent in most EC countries. The problems of participation, in particular it's apparently limited effect on health activities is essentially a problem which operates at the enterprise level. It is therefore appropriate that the social partners operating at national level and within enterprises should take responsibility for establishing effective participative structures.

8.2 An Action Plan to Encourage Health Promotion in the Workplace

In the eight member states taking part in the research, there is considerable evidence that companies are becoming involved in workplace health action. However, there are a number of problems identified from the research which largely relate to a lack of information and a lack of relevant expertise and skills concerning how to establish workplace health promotion. Against this background, it is essential that models of good practice become more widely disseminated and that programmes which help provide the knowledge, expertise and skills necessary for the establishment of workplace health promotion are fostered.

In particular, the problems of SME's need to be addressed. There is considerable evidence from the research that SME's are the least active in the area of health actions in the workplace and that many of the resource problems that they face (in particular, lack of information) are major handicaps to their becoming involved in workplace health action.

Against this background three elements of an action plan are proposed:

1. Develop a case book and database of successful health actions in the workplace.

This casebook and database should particularly focus on actions which have elements of health promotion in their structure. It should concentrate on how this health actions were established, it should outline the benefits to be gained from engaging in health actions and lastly it should illustrate how problems in their establishment were overcome. This case book should be made widely available, both in paper form, in book form and also using electronic vehicles such as remote access databases.

2. Improve cooperation between agents who have skills relevant to establishing workplace health action.

Workplace health promotion is essentially a multi-disciplinary activity which involves the skills of agents such as Health Educators, GP's, Occupational Physicians, and many others. To some degree the skills necessary for effective health promotion already exist, but rarely are they available within the confines of a single enterprise. Against this background methods to promote cooperation between enterprises and health agents who possess the relevant skills should be established.

The goal of improved co-operation between disciplines and enterprise should be made tangible through the establishment of a series of showcase examples which should take place in each of the member states. These showcase examples would provide examples of good practice, would generate material for dissemination and training development and would provide a focus for the development of information exchange. Consideration should be given to applying a uniform model of health promotion in these examples, e.g. the HWC model.

3.   Support information exchange and the organisation of conferences.

One of the most effective means of information exchange and communication of ideas are through conferences. Conferences and seminars should be organised at a local level and would involve key actors from within enterprises as well as acknowledged experts in the field.

There is also a need for conferences which are more directed at the practitioner community where knowledge and experiences can be exchanged. Conferences such as these would have the effect of overcoming many of the barriers which exist between disciplines in relation to the incorporation of the principles of health promotion to their own practice.

Possible schedule for the action plan

To implement the elements of this action plan the following schedules proposed:

A. A preparatory phase taking place in 1992/1993.

* Collection of case studies in the form of case book and preparation for incorporation into electronic data bases.

* Support the organisational sectoral and local conferences. Establish networks of cooperation between relevant professionals and enterprises.

B. Implementation phase to start in 1993:

*   Establishment of new workplace health actions which would function as models to be copied by other enterprises.

*   Publicity and dissemination of information regarding the show case examples.

*   Ongoing evaluation of the impact of the health actions taking place in the show case examples.

REFERENCES

Anderson, R.A. (1987). Health Promotion: the Concept and Application in the Workplace. In: Matheson, H. (ed.) (1987). Health Promotion in the Workplace. Scottish Health Education Group, Edinburgh.

Aravidou, K., Sarafopoulos N., Tangas,D., Tsaraklis, Z. (1990). Innovative Measures for the Promotion of Health in the Workplace: Innovative measures for the promotion of workers' health in Greece. Department of Labour, Athens.

Aravidou, K., Saraphopoulos N., Tangas, D., Tsaraklis, Z. (1991). Innovative Workplace Action for Health: Operational methods for establishing measures. Department of Labour, Athens.

BKK, (1991). European Information Centre for Health promotion. Paper to European Conference for the Promotion of health in the Workplace. Barcelona.

Clarkson, J, Blower, E., Hunter,C., Scale, I., Nutbeam,D. (1990). Overview of Innovative Workplace Action for Health in the UK. Health Promotion Authority for Wales, Cardiff.

Clarkson, J., Blower,E., Laurence, M. (1991). Innovative Workplace Action for Health in the UK: Mechanisms for establishing initiatives. Health Promotion Authority for Wales, Cardiff.

Colacino, D. and Cohen, M. (1981). The PepsiCo Approach to a Toral Health and Fitness Prograamme. In: Marshall, J. and Cooper, C. Coping with Stress at Work: Case Studies from Industry. Gower Press, Aldershot.

Dooner, B. (1990/91). A look at some of Health and Welfare Canada's health promotion activities. Health Promotion, Winter, 29, 3, 20-28.
EC 1992

Fielding, J. (1990). Paper to European Conference for the Promotion of health in the Workplace. Barcelona.

Garzi, S., and Tonelli, S. (1990). Innovative Actions for the Protection and Promotion of Health in the Workplace: The Italian Experience. University of Florence, Florence.

Garzi, S. (1991). Innovative Workplace Action for Health: Methods of launching and developing initiatives. University of Florence, Florence.

Graca, L., Faria, M. (1991). Workplace Action for Health: The situation in Portugal. Escola Nacional de Saude Publica, Lisbon.

Grundemann, R. M. (1990). Innovative Workplace Action for Health in the Netherlands. Nederlands Instituut voor Praeventieve Gezondheidszorg TNO, Leiden.

Grundemann, R., Ellis, C. (1991). Health Promoting Activities at Work in the Netherlands. Stage 2: Factors which play a part in developing and implementing the activities. Nederlands Instituut Voor Praeventieve Gezondheidszorg-TNO, Leiden.

Hauss, F. (1990). Innovative Actions Workplace Action for Health. IGES, Berlin.

Hauss, F. (1991). New Initiatives for Health in the Workplace in Germany. IGES, Berlin.

Hauss, F. (1992). Health Policy in Multinational Companies. European Foundation for the Improvement of Living and Working Conditions, Dublin. In Press.

Haglund, B.-J. Pettersonn, B. Tillgren, P. (1991). Work for Health: Briefing Book to the Sundsvall Conference on Supportive Environments, Karolinska Institure Stockholm.

HSE (1978). Health and Safety Executive, HMSO, London.

HSE (1991). Health and Safety Executive, HMSO, London.

Karasek, R. and Theorell, T. (1990). Healthy Work: Stress, Productivity and the Reconstruction of Working Life. Basic Books, New York.

Marshall, J. and Cooper, C. (1981). Coping with Stress at Work: Case Studies from Industry. Gower Press, Aldershot.

Moncada, S. (1990). Workplace Action for Health in Spain. Centre de Salut Laboral, Barcelona.

Moncada, S. (1991). Workplace Action for Health in Spain: Findings of a management survey. Centre de Salut Laboral, Barcelona.

Paoli, P. (1991). First European Survey on the Work Environment. European Foundation for the Improvement of Living and Working Conditions, Dublin.

Rantanen, J. (1990). Occupational Health Services in Europe. World Health Organisation, Copenhagen.

Sloan, R. Gruman, J. and Allegrante, J. (1987). Investing in Employee Health - a Guide to Effective Health Promotiion in the Workplace. Jossey-Bass, San Francisco.

Sworder, G. (1981). Stress Training in ICI Ltd. In: Marshall, J. and Cooper, C. Coping with Stress at Work: Case Studies from Industry. Gower Press, Aldershot.

Washington Business Group, (1987). Worksite Wellness Media Reports. Washington Business Group on Health and U.S. Department of Health and Human Services, Washington.

WHO, (1984). Health Promotion: a Discussion Document on the Concept and Principles. World Health Organisation, Copenhagen.

Wynne, R. (1989). Workplace Action for Health: A selective review and a framework for analysis. Work Research Centre, Dublin. Working Paper No. EF/WP/89/30/EN. European Foundation for the Improvement of Living and Working Conditions, Dublin.

Wynne, R. (1990). Innovative Workplace Action for Health: An overview of the Irish situation. Work Research Centre, Dublin.

Wynne, R. (1990). Innovative Workplace Actions for Health: An Overview of the Situation in Seven EC Countries. Work Research Centre, Dublin. Working Paper No. EF/WP/90/35/EN. European Foundation for the Improvement of Living and Working Conditions, Dublin.

Wynne, R. and Clarkin, N. (1991). Innovative Workplace Action for Health: Mechanisms for establishing initiatives. Work Research Centre, Dublin.

Appendix 1.

Reports from the Research Programme to date.

Hauss, F. (1992). Health Policy in Multinational Companies. European Foundation for the Improvement of Living and Working Conditions, Dublin. In Press.

Wynne, R. (1989). Workplace Action for Health: A selective review and a framework for analysis. Work Research Centre, Dublin. Working Paper No. EF/WP/89/30/EN. European Foundation for the Improvement of Living and Working Conditions, Dublin.

Wynne, R. (1990). Innovative Workplace Actions for Health: An Overview of the Situation in Seven EC Countries. Work Research Centre, Dublin. Working Paper No. EF/WP/90/35/EN. European Foundation for the Improvement of Living and Working Conditions, Dublin.

Wynne, R., and Clarkin, N. (1992). Under Construction: Building for Health in the EC Workplaces. European Foundation for the Improvement of Living and Working Conditions, Dublin. In Press.

Appendix 2.

The research questionnaire.

SECTION A : COMPANY DETAILS

1. Is your company/organisation in the ...? *(Please tick one box)*

    Public sector ☐   Private sector ☐   Other ☐ _____ *(Please specify)*

2. Is your company a multinational ...?

    Yes ☐   No ☐

3. Please describe briefly the main line of business of the company

    _____

    _____

4. What is your job title ?

    _____

5. Approximately how many employees are in your organisation ? *(Please refer only to the site at which you work)*

    Total ☐   Male ☐   Female ☐

6. Approximately what percentage of staff are members of a Trades Union? ☐

7. Please indicate the type of premises which you operate at this site *(Tick box(es) which apply)*

    Offices ☐   Production/manufacturing areas ☐   Public area, e.g. retailer ☐   Other ☐

8. Does your company provide an Occupational Health Service on site which has qualified medical or nursing staff?

    Yes ☐   No ☐

9a. Do you have a health and safety committee? (or equivalent)

    Yes ☐   No ☐   *(If No, go Question 10)*

9b Who is represented on this committee ? *(Please tick)*

    staff ☐   management ☐   unions ☐   occupational health staff ☐   other *(please specify)* ☐ _____

10.   **What do you think are the most important health problems of your employees ?**

A. _____

B. _____

C. _____

D. _____

E. _____

SECTION B : ACTIONS FOR HEALTH

This section aims to identify any actions for health which have taken place in your organisation <u>in recent years</u>. They include actions which affect the health and wellbeing of the workforce, or affect the health behaviour of the workforce, or which make the work environment more health conducive. A common feature of all of these actions is that they are supportive, i.e. they are not disciplinary procedures, but emphasise opportunities for preventive and promotive action.

Examples of health actions include: screening programmes aimed at detecting and managing hypertension, with supporting advice and information available for those who need to reduce their blood pressure; introducing healthier food choices in a workplace canteen; offering training in relaxation or psychosocial skills; and altering the design of shift schedules to maintain and improve health.

For each of the actions listed below, we would like to know:

*   If the action has taken place; and
*   The extent improving the health of the workforce was a consideration in the action.

If there are health initiatives which are not specifically listed below, please write them in the spaces provided at the end of the list.

Action	Has the action taken place ?		To what extent was improving the health of the workforce a consideration in this action?			
	Yes	No	Not at All	To some extent	To a large extent	Not Sure
Periodic executive health screening	☐	☐	☐	☐	☐	☐
Periodic health screening for all staff	☐	☐	☐	☐	☐	☐
Periodic health screening for at risk staff (e.g. for staff using dangerous chemicals, dangerous procedures)	☐	☐	☐	☐	☐	☐

Action	Has the action taken place?		To what extent was improving the health of the workforce a consideration in this action?			
	Yes	No	Not at All	To some extent	To a large extent	Not Sure
Policy on alcohol/substance misuse (N.B. not solely disciplinary in nature)	☐	☐	☐	☐	☐	☐
Introduction of a no-smoking policy/programme	☐	☐	☐	☐	☐	☐
Healthy eating policy in canteen, e.g. healthy food choices, events, promotions	☐	☐	☐	☐	☐	☐
Counselling support (i.e. personal support by trained staff for psycho-social and mental problems)	☐	☐	☐	☐	☐	☐
Exercise facilities on site/nearby	☐	☐	☐	☐	☐	☐
Exercise or lifestyle classes for staff in work time e.g. keep fit, lose weight	☐	☐	☐	☐	☐	☐
Rest facilities, social facilities, showers	☐	☐	☐	☐	☐	☐
Health education programmes, e.g. posters, displays, leaflets, videos, staff newsletters	☐	☐	☐	☐	☐	☐
Redesign of shift schedules	☐	☐	☐	☐	☐	☐
Stress control programme	☐	☐	☐	☐	☐	☐
Job design programmes, i.e. designing individual jobs	☐	☐	☐	☐	☐	☐
Work organisation programmes, i.e. designing how jobs fit together	☐	☐	☐	☐	☐	☐
Flexibilisation of working time, e.g. rest breaks, holidays	☐	☐	☐	☐	☐	☐
Welfare support programmes	☐	☐	☐	☐	☐	☐
Support programmes (non-care), e.g. Alcoholics/Gamblers Anonymous	☐	☐	☐	☐	☐	☐
Training in Human Resource Management skills	☐	☐	☐	☐	☐	☐
Community and social programmes	☐	☐	☐	☐	☐	☐

Action	Has the action taken place ?		To what extent was improving the health of the workforce a consideration in this action?			
	Yes	No	Not at All	To some extent	To a large extent	Not Sure
Toxic substance control	☐	☐	☐	☐	☐	☐
Introducing guards on machinery	☐	☐	☐	☐	☐	☐
Protective clothing and equipment	☐	☐	☐	☐	☐	☐
Automating hazardous processes	☐	☐	☐	☐	☐	☐
Changes to any of the following aspects of the physical environment						
- individual workspaces	☐	☐	☐	☐	☐	☐
- lighting	☐	☐	☐	☐	☐	☐
- heating/air conditioning	☐	☐	☐	☐	☐	☐
- ventilation	☐	☐	☐	☐	☐	☐
- Interior design	☐	☐	☐	☐	☐	☐
- Noise reduction	☐	☐	☐	☐	☐	☐
Any other actions to improve the health of workers						
_____	☐	☐	☐	☐	☐	☐
_____	☐	☐	☐	☐	☐	☐
_____	☐	☐	☐	☐	☐	☐

Have you any other comments to make about actions for health in your organisation ?

_____

_____

_____

_____

_____

_____

SECTION C : ESTABLISHING ACTION FOR HEALTH

Here we are interested in the process of how health actions became established in your organisation. The following questions ask what factors prompted action, whether the workforce was consulted or involved, and who is responsible for implementing action. There is some space allocated at the end of the section for you to write in any additional information which you feel is relevant.

1. To what extent did any of the following factors prompt any of the health actions in your company ?

	Not at All	To some extent	To a great extent	Not Sure
Legislation	☐	☐	☐	☐
Personnel/welfare problems	☐	☐	☐	☐
Health problems in the workforce	☐	☐	☐	☐
Staff morale	☐	☐	☐	☐
Absenteeism	☐	☐	☐	☐
Productivity/Performance	☐	☐	☐	☐
Staff turnover	☐	☐	☐	☐
Industrial relations problems	☐	☐	☐	☐
Company public image	☐	☐	☐	☐
Accident rates	☐	☐	☐	☐
Other reasons (please specify)				
_____	☐	☐	☐	☐
_____	☐	☐	☐	☐
_____	☐	☐	☐	☐

2. **Health actions can follow a lifecycle involving: the initial idea, planning, implementation and monitoring/evaluation. Are any of the following people usually involved in these stages of health actions in your organisation ?** *(If a stage does not take place, please tick no for each type of person)*

	Initial idea Yes No	Planning Yes No	Implementation Yes No	Evaluation Yes No
Management	☐ ☐	☐ ☐	☐ ☐	☐ ☐
Staff representatives	☐ ☐	☐ ☐	☐ ☐	☐ ☐
Trade Union reps.	☐ ☐	☐ ☐	☐ ☐	☐ ☐
Health and safety reps.	☐ ☐	☐ ☐	☐ ☐	☐ ☐
Occupational health staff	☐ ☐	☐ ☐	☐ ☐	☐ ☐
Outside consultants	☐ ☐	☐ ☐	☐ ☐	☐ ☐
Others (*Please specify*) _____	☐ ☐	☐ ☐	☐ ☐	☐ ☐
_____	☐ ☐	☐ ☐	☐ ☐	☐ ☐

3. **Being involved in health actions can take many forms. For example, people can be given information about an action, can be consulted about an action, can participate in decisions about the action or can be responsible for planning the action. Please indicate below the usual type of involvement, during this PLANNING PHASE of health actions, for different staff in your organisation.**

N.B. Planning Phase	Information Yes No	Consultation Yes No	Participation Yes No	Responsibility Yes No
Management	☐ ☐	☐ ☐	☐ ☐	☐ ☐
Staff representatives	☐ ☐	☐ ☐	☐ ☐	☐ ☐
Trade Union reps.	☐ ☐	☐ ☐	☐ ☐	☐ ☐
Health and safety reps.	☐ ☐	☐ ☐	☐ ☐	☐ ☐
Occupational health staff	☐ ☐	☐ ☐	☐ ☐	☐ ☐
Outside consultants	☐ ☐	☐ ☐	☐ ☐	☐ ☐
Others (*Please specify*) _____	☐ ☐	☐ ☐	☐ ☐	☐ ☐
_____	☐ ☐	☐ ☐	☐ ☐	☐ ☐

4. **The process of immpementing health actions can take many forms. For example, people can be given <u>information</u> about an action, can be <u>consulted</u> about an action, can <u>participate</u> in implementing the action or can be <u>responsible</u> for implementing the action. Please indicate below the usual type of involvement during the IMPLEMENTATION PHASE of health actions for different staff in your organisation.**

N.B. Implementation Phase	Information Yes No	Consultation Yes No	Participation Yes No	Responsibility Yes No
Management	☐ ☐	☐ ☐	☐ ☐	☐ ☐
Staff representatives	☐ ☐	☐ ☐	☐ ☐	☐ ☐
Trade Union reps.	☐ ☐	☐ ☐	☐ ☐	☐ ☐
Health and safety reps.	☐ ☐	☐ ☐	☐ ☐	☐ ☐
Occupational health staff	☐ ☐	☐ ☐	☐ ☐	☐ ☐
Outside consultants	☐ ☐	☐ ☐	☐ ☐	☐ ☐
Others (*Please specify*) _____ _____	☐ ☐ ☐ ☐	☐ ☐ ☐ ☐	☐ ☐ ☐ ☐	☐ ☐ ☐ ☐

5. **What are the three most important barriers to actions for health being established in your organisation ?**

A. _____

B. _____

C. _____

6. **To what extent has your organisation encountered the following kinds of problems in establishing health actions ?**

	Not at All	To some extent	To a great extent	Not Sure
Lack of human resources	☐	☐	☐	☐
Lack of financial resources	☐	☐	☐	☐
Lack of material resources	☐	☐	☐	☐
Lack of suitable facilities	☐	☐	☐	☐
Lack of management commitment	☐	☐	☐	☐
Lack of worker commitment	☐	☐	☐	☐
Lack of expertise	☐	☐	☐	☐
Difficulty of achieving consensus	☐	☐	☐	☐
Others *(Please specify)*	☐	☐	☐	☐
_____	☐	☐	☐	☐
_____	☐	☐	☐	☐

7. **What were the strategies your organisation took to to overcome these problems that you believe were successful ?**

Problem	Action to overcome problem

8. **What are the three most important benefits your organisation has observed as a result of taking actions for health ?**

A. _____

B. _____

C. _____

9. **To what extent have the following benefits have been observed in your organisation as a result of actions for health ?**

	Not at All	To some extent	To a great extent	Not Sure
Reduced personnel/welfare problems	☐	☐	☐	☐
Improved health in the workforce	☐	☐	☐	☐
Improved staff morale/climate	☐	☐	☐	☐
Reduced absenteeism	☐	☐	☐	☐
Improved productivity/performance	☐	☐	☐	☐
Reduced staff turnover	☐	☐	☐	☐
Improved industrial relations	☐	☐	☐	☐
Improved company public image	☐	☐	☐	☐
Reduced accident rates	☐	☐	☐	☐
Others *(Please specify)*				
_____	☐	☐	☐	☐
_____	☐	☐	☐	☐
_____	☐	☐	☐	☐

SECTION D : PLANS AND PRIORITIES

This section is concerned with your views on the level of support your organisation offers staff to enable them lead healthy lives, or whether the organisation could do more. We also enquire about your organisation's plans for initiatives in the near future.

1. **In your opinion, on a scale of 1-10, where would you say that the health of the workforce should lie in terms of the priorities of the organisation? (For example in comparison with productivity or wage settlements).** *(Please circle one number).*

    Highest Priority   1   2   3   4   5   6   7   8   9   10   Lowest Priority

2. **In your opinion, on a scale of 1-10, where would you say that your organisation places the health of the workforce in terms of priorities?**

    Highest Priority   1   2   3   4   5   6   7   8   9   10   Lowest Priority

3. **Does your company have an explicit or written health policy ?** *(We are not referring to a safety policy alone here)*

   Yes ☐   No *(If No, go to question 4)* ☐

3a. **What is this policy ?**

   _____

   _____

   _____

   _____

   _____

3b **Which groups have been involved in the development of this policy ?**

	Yes	No		Yes	No
Management	☐	☐	Health and safety reps.	☐	☐
Occupational health staff	☐	☐	Outside consultants	☐	☐
Staff representatives	☐	☐	Others _____	☐	☐
Trade Union reps.	☐	☐			

4. **Does your company plan to further develop a health policy ?**

   Yes ☐   No ☐

5. **Does your organisation have a budget specifically set aside for health (<u>not safety</u>) actions in the workplace ?**

   Yes ☐   No ☐

6. **Approximately what percentage of company turnover is spent on health (not safety) in the workplace (including salaries for occupational health staff, capital expenditure, expenditure on specific programmes) ?**

   Percentage ☐   Don't know ☐   Amount ☐

6a. **Will this proportion increase or decrease in the future ?**

   Increase ☐   Decrease ☐   Stay the same ☐

7. Finally, if your organisation is planning to introduce any new actions for health within the next 12 months, please describe briefly what actions are planned.

    _____
    _____
    _____
    _____
    _____

8. Have you any further comments you would like to make?

    _____
    _____
    _____
    _____
    _____

Thank you for completing this questionnaire

You may be interested to know what the information you have provided will be used for. This questionnaire is part of a survey in EC countries which seeks to:

* assess the nature and extent of workplace health actions;
* identify the barriers to establishing health actions in the workplace
* identify examples of good practice so that they can be disseminated widely
* compare experiences between regions and between countries;
* promote awareness of how to establish health actions in the workplace

Finally, in connection with this research sponsored by the European Foundation for the Improvement of Living and Working Conditions...

	Yes	No
May we name your company in any reports arising from the research ?	☐	☐
Would you like to participate in information seminars resulting from the research ?	☐	☐
Would you like results from the research ?	☐	☐

European Foundation for the Improvement of Living and Working Conditions

UNDER CONSTRUCTION
Building for Health in the EC Workplace

Luxembourg: Office for Official Publications of the European Communities

1992 — 217 p. — 23.5 × 16 cm

ISBN 92-826-4629-7

Price (excluding VAT) in Luxembourg ECU 13.50

Venta y suscripciones • Salg og abonnement • Verkauf und Abonnement • Πωλήσεις και συνδρομές • Sales and subscriptions • Vente et abonnements • Vendita e abbonamenti • Verkoop en abonnementen • Venda e assinaturas

BELGIQUE / BELGIË

**Moniteur belge /
Belgisch Staatsblad**
Rue de Louvain 42 / Leuvenseweg 42
1000 Bruxelles / 1000 Brussel
Tél. (02) 512 00 26
Fax 511 01 84
CCP / Postrekening 000-2005502-27

Autres distributeurs /
Overige verkooppunten

**Librairie européenne/
Europese Boekhandel**
Avenue Albert Jonnart 50 /
Albert Jonnartlaan 50
1200 Bruxelles / 1200 Brussel
Tél. (02) 734 02 81
Fax 735 08 60

Jean De Lannoy
Avenue du Roi 202 /Koningslaan 202
1060 Bruxelles / 1060 Brussel
Tél. (02) 538 51 69
Télex 63220 UNBOOK B
Fax (02) 538 08 41

CREDOC
Rue de la Montagne 34 / Bergstraat 34
Bte 11 / Bus 11
1000 Bruxelles / 1000 Brussel

DANMARK

**J. H. Schultz Information A/S
EF-Publikationer**
Ottiliavej 18
2500 Valby
Tlf. 36 44 22 66
Fax 36 44 01 41
Girokonto 6 00 08 86

BR DEUTSCHLAND

Bundesanzeiger Verlag
Breite Straße
Postfach 10 80 06
5000 Köln 1
Tel. (02 21) 20 29-0
Telex ANZEIGER BONN 8 882 595
Fax 20 29 278

GREECE/ΕΛΛΑΔΑ

G.C. Eleftheroudakis SA
International Bookstore
Nikis Street 4
10563 Athens
Tel. (01) 322 63 23
Telex 219410 ELEF
Fax 323 98 21

ESPAÑA

Boletín Oficial del Estado
Trafalgar, 27
28010 Madrid
Tel. (91) 44 82 135

Mundi-Prensa Libros, S.A.
Castelló, 37
28001 Madrid
Tel. (91) 431 33 99 (Libros)
     431 32 22 (Suscripciones)
     435 36 37 (Dirección)
Télex 49370-MPLI-E
Fax (91) 575 39 98

Sucursal:

Librería Internacional AEDOS
Consejo de Ciento, 391
08009 Barcelona
Tel. (93) 301 86 15
Fax (93) 317 01 41

**Llibreria de la Generalitat
de Catalunya**
Rambla dels Estudis, 118 (Palau Moja)
08002 Barcelona
Tel. (93) 302 68 35
     302 64 62
Fax (93) 302 12 99

FRANCE

**Journal officiel
Service des publications
des Communautés européennes**
26, rue Desaix
75727 Paris Cedex 15
Tél. (1) 40 58 75 00
Fax (1) 40 58 75 74

IRELAND

Government Supplies Agency
4-5 Harcourt Road
Dublin 2
Tel. (1) 61 31 11
Fax (1) 78 06 45

ITALIA

Licosa Spa
Via Duca di Calabria, 1/1
Casella postale 552
50125 Firenze
Tel. (055) 64 54 15
Fax 64 12 57
Telex 570466 LICOSA I
CCP 343 509

GRAND-DUCHÉ DE LUXEMBOURG

Messageries Paul Kraus
11, rue Christophe Plantin
2339 Luxembourg
Tél. 499 88 88
Télex 2515
Fax 499 88 84 44
CCP 49242-63

NEDERLAND

SDU Overheidsinformatie
Externe Fondsen
Postbus 20014
2500 EA 's-Gravenhage
Tel. (070) 37 89 911
Fax (070) 34 75 778

PORTUGAL

Imprensa Nacional
Casa da Moeda, EP
Rua D. Francisco Manuel de Melo, 5
1092 Lisboa Codex
Tel. (01) 69 34 14

**Distribuidora de Livros
Bertrand, Ld.ª**
Grupo Bertrand, SA
Rua das Terras dos Vales, 4-A
Apartado 37
2700 Amadora Codex
Tel. (01) 49 59 050
Telex 15798 BERDIS
Fax 49 60 255

UNITED KINGDOM

HMSO Books (PC 16)
HMSO Publications Centre
51 Nine Elms Lane
London SW8 5DR
Tel. (071) 873 2000
Fax GP3 873 8463
Telex 29 71 138

ÖSTERREICH

**Manz'sche Verlags-
und Universitätsbuchhandlung**
Kohlmarkt 16
1014 Wien
Tel. (0222) 531 61-0
Telex 11 25 00 BOX A
Fax (0222) 531 61-39

SUOMI

Akateeminen Kirjakauppa
Keskuskatu 1
PO Box 128
00101 Helsinki
Tel. (0) 121 41
Fax (0) 121 44 41

NORGE

Narvesen information center
Bertrand Narvesens vei 2
PO Box 6125 Etterstad
0602 Oslo 6
Tel. (2) 57 33 00
Telex 79668 NIC N
Fax (2) 68 19 01

SVERIGE

BTJ
Box 200
22100 Lund
Tel. (046) 18 00 00
Fax (046) 18 01 25

SCHWEIZ / SUISSE / SVIZZERA

OSEC
Stampfenbachstraße 85
8035 Zürich
Tel. (01) 365 54 49
Fax (01) 365 54 11

CESKOSLOVENSKO

NIS
Havelkova 22
13000 Praha 3
Tel. (02) 235 84 46
Fax 42-2-264775

MAGYARORSZÁG

Euro-Info-Service
Budapest I. Kir.
Attila út 93
1012 Budapest
Tel. (1) 56 82 11
Telex (22) 4717 AGINF H-61
Fax (1) 17 59 031

POLSKA

Business Foundation
ul. Krucza 38/42
00-512 Warszawa
Tel. (22) 21 99 93, 628-28-82
International Fax&Phone
(0-39) 12-00-77

JUGOSLAVIJA

Privredni Vjesnik
Bulevar Lenjina 171/XIV
11070 Beograd
Tel. (11) 123 23 40

CYPRUS

Cyprus Chamber of Commerce and Industry
Chamber Building
38 Grivas Dhigenis Ave
3 Deligiorgis Street
PO Box 1455
Nicosia
Tel. (2) 449500/462312
Fax (2) 458630

TÜRKIYE

**Pres Gazete Kitap Dergi
Pazarlama Dağitim Ticaret ve sanayi
AŞ**
Narlibahçe Sokak N. 15
Istanbul-Cağaloğlu
Tel. (1) 520 92 96 - 528 55 66
Fax 520 64 57
Telex 23822 DSVO-TR

CANADA

Renouf Publishing Co. Ltd
Mail orders — Head Office:
1294 Algoma Road
Ottawa, Ontario K1B 3W8
Tel. (613) 741 43 33
Fax (613) 741 54 39
Telex 0534783

Ottawa Store:
61 Sparks Street
Tel. (613) 238 89 85

Toronto Store:
211 Yonge Street
Tel. (416) 363 31 71

UNITED STATES OF AMERICA

UNIPUB
4611-F Assembly Drive
Lanham, MD 20706-4391
Tel. Toll Free (800) 274 4888
Fax (301) 459 0056

AUSTRALIA

Hunter Publications
58A Gipps Street
Collingwood
Victoria 3066

JAPAN

Kinokuniya Company Ltd
17-7 Shinjuku 3-Chome
Shinjuku-ku
Tokyo 160-91
Tel. (03) 3439-0121

Journal Department
PO Box 55 Chitose
Tokyo 156
Tel. (03) 3439-0124

AUTRES PAYS
OTHER COUNTRIES
ANDERE LÄNDER

**Office des publications officielles
des Communautés européennes**
2, rue Mercier
2985 Luxembourg
Tél. 49 92 81
Télex PUBOF LU 1324 b
Fax 48 85 73/48 68 17
CC bancaire BIL 8-109/6003/700